Deploy Machine Learning Models to Production

With Flask, Streamlit, Docker, and Kubernetes on Google Cloud Platform

Pramod Singh

Apress®

Deploy Machine Learning Models to Production

Pramod Singh
Bangalore, Karnataka, India

ISBN-13 (pbk): 978-1-4842-6545-1   ISBN-13 (electronic): 978-1-4842-6546-8
https://doi.org/10.1007/978-1-4842-6546-8

Managing Director, Apress Media LLC: Welmoed Spahr
Acquisitions Editor: Celestin Suresh John
Development Editor: Laura Berendson
Coordinating Editor: Aditee Mirashi

Cover designed by eStudioCalamar

Cover image designed by Freepik (www.freepik.com)

Distributed to the book trade worldwide by Springer Science+Business Media New York, 1 New York Plaza, Suite 4600, New York, NY 10004-1562, USA. Phone 1-800-SPRINGER, fax (201) 348-4505, email orders-ny@springer-sbm.com, or visit www.springeronline.com. Apress Media, LLC is a California LLC and the sole member (owner) is Springer Science + Business Media Finance Inc (SSBM Finance Inc). SSBM Finance Inc is a **Delaware** corporation.

For information on translations, please e-mail booktranslations@springernature.com; for reprint, paperback, or audio rights, please e-mail bookpermissions@springernature.com.

Apress titles may be purchased in bulk for academic, corporate, or promotional use. eBook versions and licenses are also available for most titles. For more information, reference our Print and eBook Bulk Sales web page at www.apress.com/bulk-sales.

Any source code or other supplementary material referenced by the author in this book is available to readers on GitHub via the book's product page, located at www.apress.com/ 978-1-4842-6545-1. For more detailed information, please visit www.apress.com/source-code.

Printed on acid-free paper

Table of Contents

About the Author

Pramod Singh is a manager of data science at Bain & Company. He has more than 11 years of rich experience in the data science field working with multiple product- and service-based organizations. He has been part of numerous large-scale ML and AI projects. He has published three books on large-scale data processing and machine learning. He is also a regular speaker at major AI conferences such as O'Reilly AI and Strata.

About the Technical Reviewer

 **Manohar Swamynathan** is a data science practitioner and an avid programmer, with 14+ years of experience in various data science areas that include data warehousing, business intelligence (BI), analytical tool development, ad hoc analysis, predictive modeling, data science product development, consulting, formulating strategy, and executing analytics programs. He's had a career covering the life cycle of data across different domains such as US mortgage banking, retail/e-commerce, insurance, and industrial IoT. He has a bachelor's degree with a specialization in physics, mathematics, and computers, and a master's degree in project management. He's currently living in Bengaluru, the Silicon Valley of India.

He has also been the technical reviewer of books such as *Data Science Using Python and R.*

Acknowledgments

I want to take a moment to thank the most important person in my life: my wife, Neha. Without her support, this book wouldn't have seen the light of day. She is the source of my energy, motivation, and happiness and keeps me going despite challenges and hardships. I dedicate this book to her.

I also want to thank a few other people who helped a great deal during these months and provided a lot of support. Let me start with Aditee, who was very patient and kind to understand the situation and help to reorganize the schedule. Thanks to Celestian John as well to offer me another opportunity to write for Apress. Last but not the least, my mentors: Barron Beranjan, Janani Sriram, Sebastian Keupers, Sreenivas Venkatraman, Dr. Vijay Agneeswaran, Shoaib Ahmed, and Abhishek Kumar. Thank you for your continuous guidance and support.

Introduction

This book helps upcoming data scientists who have never deployed any machine learning model. Most data scientists spend a lot of time analyzing data and building models in Jupyter Notebooks but have never gotten an opportunity to take them to the next level where those ML models are exposed as APIs. This book helps those people in particular who want to deploy these ML models in production and use the power of these models in the background of a running application.

The term *ML productionization* covers lots of components and platforms. The core idea of this book is not to look at each of the options available but rather provide a holistic view on the frameworks for productionizing models, from basic ML-based apps to complex ones. Once you know how to take an ML model and put it in production, you will become more confident to work on complicated applications and big deployments. This book covers different options to expose the ML model as a web service using frameworks such as Flask and Streamlit. It also helps readers to understand the usage of Docker in machine learning apps and the end-to-end process of deployment on Google Cloud Platform using Kubernetes.

I hope there is some useful information for every reader, and potentially they can apply it in their workstreams to go beyond Jupyter Notebooks and productionalize some of their ML models.

CHAPTER 1

Introduction to Machine Learning

In this first chapter, we are going to discuss some of the fundamentals of machine learning and deep learning. We are also going to look at different business verticals that are being transformed by using machine learning. Finally, we are going to go over the traditional steps of training and building a rather simple machine learning model and deep learning model on a cloud platform (Databricks) before moving on to the next set of chapters on productionization. If you are aware of these concepts and feel comfortable with your level of expertise on machine learning already, I encourage you to skip the next two sections and move on to the last section, where I mention the development environment and give pointers to the book's accompanying codebase and data download information so that you are able to set up the environment appropriately. This chapter is divided into three sections. The first section covers the introduction to the fundamentals of machine learning. The second section dives into the basics of deep learning and the details of widely used deep neural networks. Each of the previous sections is followed up by the code to build a model on the cloud platform. The final section is about the requirements and environment setup for the remainder of the chapters in the book.

© Pramod Singh 2021
P. Singh, *Deploy Machine Learning Models to Production*,
https://doi.org/10.1007/978-1-4842-6546-8_1

History

Machine learning/deep learning is not new; in fact, it goes back to 1940s when for the first time an attempt was made to build something that had some amount of built-in intelligence. The great Alan Turing worked on building this unique machine that could decrypt German code during World War II. That was the beginning of machine intelligence era, and within a few years, researchers started exploring this field in great detail across many countries. ML/DL was considered to be significantly powerful in terms of transforming the world at that time, and an enormous number of funds were granted to bring it to life. Nearly everybody was very optimistic. By late 1960s, people were already working on machine vision learning and developing robots with machine intelligence.

While it all looked good on the surface level, there were some serious challenges that were impeding the progress in this field. Researchers were finding it extremely difficult to create intelligence in the machines. Primarily it was due to a couple of reasons. One of them was that the processing power of computers in those days was not enough to handle and process large amounts of data, and the reason was the availability of relevant data itself. Despite the support of government and the availability of sufficient funds, the ML/AI research hit a roadblock from the period of the late 1960s to the early 1990s. This block of time period is also known as the "AI winters" among the community members.

In the late 1990s, corporations once again became interested in AI. The Japanese government unveiled plans to develop a fifth-generation computer to advance machine learning. AI enthusiasts believed that soon computers would be able to carry on conversations, translate languages, interpret pictures, and reason like people. In 1997, IBM's Deep Blue became the first computer to beat a reigning world chess champion, Garry Kasparov. Some AI funding dried up when the dot-com bubble burst in the early 2000s. Yet machine learning continued its march, largely thanks to improvements in computer hardware.

The Last Decade

There is no denying the fact that the world has seen significant progress in terms of machine learning and AI applications in the last decade or so. In fact, if it were to be compared with any other technology, ML/AI has been path-breaking in multiple ways. Businesses such as Amazon, Google, and Facebook are thriving on these advancements in AI and are partly responsible for it as well. The research and development wings of organizations like these are pushing the limits and making incredible progress in bringing AI to everyone. Not only big names like these but thousands of startups have emerged on the landscape specializing in AI-based products and services. The numbers only continue to grow as I write this chapter. As mentioned earlier, the adoption of ML and AI by various businesses has exponentially grown over the last decade or so, and the prime reason for this behavior has been multifold.

- Rise in data

- Increased computational efficiency

- Improved ML algorithms

- Availability of data scientists

Rise in Data

The first most prominent reason for this trend is the massive rise in data generation in the past couple of decades. Data was always present, but it's imperative to understand the exact reason behind this abundance of data. In the early days, the data was generated by employees or workers of particular organizations as they would save the data into systems, but there were limited data points holding only a few variables. Then came the revolutionary Internet, and generic information was made accessible to virtually everyone using the Internet. With the Internet, the users got

3

the control to enter and generate their own data. This was a colossal shift as the total number of Internet users in the world grew at an exploding rate, and the amount of data created by these users grew at an even higher rate. All of this data—login/sign-up forms capturing user details, photos and videos uploads on various social platforms, and other online activities—led to the coining of the term Big Data. As a result, the challenges that ML and AI researchers faced in earlier times due to a lack of data points were completely eliminated, and this proved to be a major enabler for the adoption of in ML and AI.

Finally, from a data perspective, we have already reached the next level as machines are generating and accumulating data. Every device around us is capturing data such as cars, buildings, mobiles, watches, and flight engines. They are embedded with multiple monitoring sensors and are recording data every second. This data is even higher in magnitude than the user-generated data and commonly referred as Internet of Things (IoT) data.

Increased Computational Efficiency

We have to understand the fact that ML and AI at the end of the day are simply dealing with a huge set of numbers being put together and made sense out of. To apply ML or AI, there is a heavy need for powerful processing systems, and we have witnessed significant improvements in computation power at a breakneck pace. Just to observe the changes that we have seen in the last decade or so, the size of mobile devices has reduced drastically, and the speed has increased to a great extent. This is not just in terms of physical changes in the microprocessor chips for faster processing using GPUs and TPUs but also in the presence of data processing frameworks such as Spark. The combination of advancement in processing capabilities and in-memory computations using Spark made it possible for lots of ML algorithms to be able to run successfully in the past decade.

Improved ML Algorithms

Over the last few years, there has been tremendous progress in terms of the availability of new and upgraded algorithms that have not only improved the predictions accuracy but also solved multiple challenges that traditional ML faced. In the first phase, which was a rule-based system, one had to define all the rules first and then design the system within those set of rules. It became increasingly difficult to control and update the number of rules as the environment was too dynamic. Hence, traditional ML came into the picture to replace rule-based systems. The challenge with this approach was that the data scientist had to spent a lot of time to hand design the features for building the model (known as *feature engineering*), and there was an upper threshold in terms of predictions accuracy that these models could never go above no matter if the input data size increased. The third phase was the introduction of deep neural networks where the network would figure out the most important features on its own and also outperform other ML algorithms. In addition, some other approaches that have been creating a lot of buzz over the last few years are as follows:

- Meta learning

- Transfer learning (nano nets)

- Capsule networks

- Deep reinforcement learning

- Generative adversarial networks (GANs)

Availability of Data Scientists

ML/AI is a specialized field as the skills required to be able to do this is indeed a combination of multiple disciplines. To be able to build and apply ML models, one needs to have a sound knowledge of math and statistics fundamentals. Along with that, a deep understanding of machine learning algorithms and various optimization techniques is critical to taking the right approach to solve a business problem using ML and AI. The next important skill is to be extremely comfortable at coding, and the last one is to be an expert of particular domain (finance, retail, auto, healthcare, etc.) or carry deep knowledge of multiple domains. There is a huge excitement in the job markets with respect to data scientist roles, and there are a huge number of requirements for data scientists everywhere, especially in countries such as the United States, United Kingdom, and India.

Machine Learning

Now that we know a little bit of history around machine learning, we can go over the fundamentals of machine learning. We can break down ML into four parts, as shown in Figure 1-1.

- Supervised machine learning

- Unsupervised machine learning

- Semi-supervised machine learning

- Reinforcement machine learning

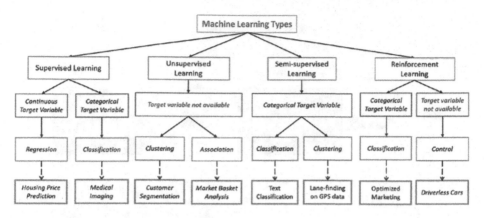

Figure 1-1. *Machine learning categories (source: en.proft.me)*

Supervised Machine Learning

Supervised machine learning is the major category of machine learning that drives a lot of applications and value for businesses. In this type of learning, the model is trained on the data for which we already have the correct labels or output. In short, we try to map the relationship between input data and output data in such a way that it can generalize well on unseen data as well, as shown in Figure 1-2. The training of the model takes place by comparing the actual output with the predicted output and then optimizing the function to reduce the total error between the actual and predicted.

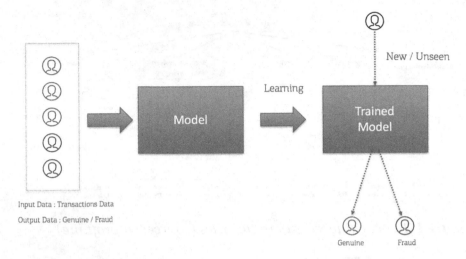

Figure 1-2. *Generalization*

This type of learning is predominantly used in cases where historical data is available and predictions need to be made on future data. The further categorization of supervised learning is based on types of labels being used for prediction, as shown in Figure 1-3. If the nature of the output variable is numerical, it falls under regression, whereas if it is categorical, it is in the classification category.

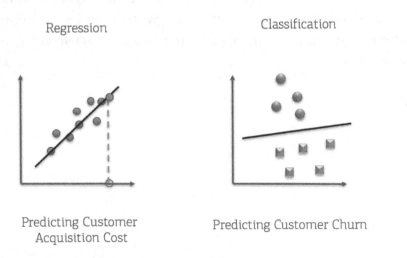

Figure 1-3. *Regression versus classification*

Classification refers to the case when the output variable is a discrete value or categorical in nature. Classification comes in two types.

- Binary classification

- Multiclassification

When the target class is of two categories, it is referred to as *binary*, and when it is more than two classes, it is known as multiclassifications, as shown in Figure 1-4.

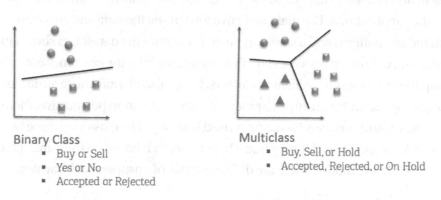

Binary Class
- Buy or Sell
- Yes or No
- Accepted or Rejected

Multiclass
- Buy, Sell, or Hold
- Accepted, Rejected, or On Hold

Figure 1-4. *Binary versus multiclass*

Another property of supervised learning is that the model's performance can be evaluated. Based on the type of model (classification or regression), the evaluation metric can be applied, and performance results can be measured. This happens mainly by splitting the training data into two sets (the train set and the validation set) and training the model on the train set and testing its performance on the validation set since we already know the right label/outcome for the validation set.

Unsupervised Learning

Unsupervised learning is another category of machine learning that is used heavily in business applications. It is different from supervised learning in terms of the output labels. In unsupervised learning, we build the models on similar sort of data as of supervised learning except for the fact that this dataset does not contain any label or outcomes column. Essentially, we apply the model on the data without any right answers. In unsupervised learning, the machine tries to find hidden patterns and useful signals in the data that can be later used for other applications. The main objective is to probe the data and come up with hidden patterns and a similarity structure within the dataset, as shown in Figure 1-5. One of the use cases is to find patterns within the customer data and group the customers into different clusters. It can also identify those attributes that distinguish between any two groups. From a validation perspective, there is no measure of accuracy for unsupervised learning. The clustering done by person A can be totally different from that of person B based on the parameters used to build the model. There are different types of unsupervised learning.

- K-means clustering

- Mapping of nearest neighbor

Clustering

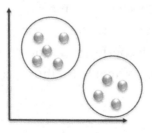

Customer Segmentation

Figure 1-5. *Clustering*

Semi-supervised Learning

As the name suggests, semi-supervised learning lies somewhere in between supervised and unsupervised learning. In fact, it uses both of the techniques. This type of learning is mainly relevant in scenarios when we are dealing with a mixed sort of dataset, which contains both labeled and unlabeled data. Sometimes it's just unlabeled data completely, but we label some part of it manually. The whole idea of semi-supervised learning is to use this small portion of labeled data to train the model and then use it for labeling the other remaining part of data, which can then be used for other purposes. This is also known as *pseudo-labeling* as it labels the unlabeled data using the predictions made by the supervised model. To quote a simple example, say we have lots of images of different brands from social media and most of it is unlabeled. Now using semi-supervised learning, we can label some of these images manually and then train our model on the labeled images. We then use the model predictions to label the remaining images to transform the unlabeled data to labeled data completely.

The next step in semi-supervised learning is to re-train the model on entire labeled dataset. The advantage that it offers is that the model gets trained on a bigger dataset, which was not the case earlier and is now more robust and better at predictions. The other advantage is that semi-supervised learning saves a lot of effort and time that could go in to manually label the data. The flipside of doing all this is that it's difficult to get the high performance of the pseudo-labeling as it uses a small part of the labeled data to make the predictions. However, it is still a better option rather than manually labeling the data, which can be expensive and time-consuming at the same time. This is how semi-supervised learning uses both the supervised and unsupervised learning to generate the labeled data. Businesses that face challenges regarding costs associated with the labeled training process usually go for semi-supervised learning.

Reinforcement Learning

Reinforcement learning is the fourth kind of learning and is little different in terms of the data usage and its predictions. Reinforcement learning is a big research area in itself, and an entire book could be written just on it. The main difference between the other kinds of learning and reinforcement learning is that we need data, mainly historical data, to train the models, whereas reinforcement learning works on a reward system, as shown in Figure 1-6. It is primarily decision-making based on certain actions that the agent takes to change its state while trying to maximize the rewards. Let's break this down to individual elements using a visualization.

Figure 1-6. *Reinforcement learning*

- *Autonomous agent*: This is the main character in this whole learning who is responsible for taking action. If it is a game, the agent makes the moves to finish or reach the end goal.

- *Actions*: These are set of possible steps that the agent can take to move forward in the task. Each action will have some effect on the state of the agent and can result in either reward or penalty. For example, in a game of tennis, the actions might be to serve, return, move left or right, etc.

- *Reward*: This is the key to making progress in reinforcement learning. Rewards enable the agents to take actions based on if they're positive rewards or penalties. It is an instant feedback mechanism that differentiates it from traditional supervised and unsupervised learning techniques.

- *Environment*: This is the territory in which the agent gets to play in. The environment decides whether the actions that the agent takes results in rewards or penalties.

- *State*: The position the agent is in at any given point of time defines the state of the agent. To move forward or reach the end goal, the agent has to keep changing states in the positive direction to maximize the rewards.

The unique thing about reinforcement learning is that there is an immediate feedback mechanism that drives the next behavior of the agent based on a reward system. Most of the applications that use reinforcement learning are in navigation, robotics, and gaming. However, it can be also used to build recommender systems.

Now let's go over some of the important concepts in machine learning as its critical to have a good understanding of these aspects before moving on to the machine learning in production.

Gradient Descent

At the end of the day, the machine learning model is as good as the loss it's able to minimize in its predictions. There are different types of loss functions pertaining to a specific category of problems, and most often in the typical classification or regression tasks, we try to minimize the mean squared error and log loss during training and cross validation. If we think of the loss as a curve, as shown in Figure 1-7, gradient descent helps us to reach the point where the loss value is at its minimum. We start a random point based on the initial weights or parameters in the model and move in the direction where it starts reducing. One thing worth remembering here is that gradient descent takes big steps when it's far away from the actual minima, whereas once it reaches a nearby value, the step sizes become very small to not miss the minima.

To move toward the minimum value point, it starts with taking the derivative of the error with respect to the parameters/coefficients (weights in case of neural networks) and tries to find the point where the slope of this error curve is equal to zero. One of the important components in gradient descent is the learning rate as it decides how quickly or how slowly it descends toward the lowest error value. If learning rate parameters are set to be higher value, then chances are that it might skip the lowest value, and on the contrary, if learning rate is too small, it would take a long time to converge. Hence, the learning rate becomes an important part in the overall gradient descent process.

The overall aim of gradient descent is to reach to a corresponding combination of input coefficients that reflect the minimum errors based on the training data. So, in a way we try to change these coefficient values from earlier values to have minimum loss. This is achieved by the process of subtracting the product of the learning rate and the slope (derivative of error with regard to the coefficient) from the old coefficient value. This alteration in coefficient values keeps happening until there is no more change in the coefficient/weights of the model as it signifies that the gradient descent has reached the minimum value point in the loss curve.

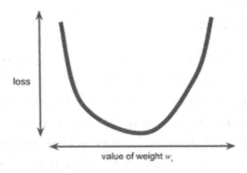

Figure 1-7. *Gradient descent*

Another type of gradient descent technique is stochastic gradient descent (SGD), which deals with a similar approach for minimizing the error toward zero but with sets of data points instead of considering all data in one go. It takes sample data from input data and applies gradient descent to find the point of lowest error.

Bias vs. Variance

Bias variance trade-off is the most common problem that gets attention from data scientists. High bias refers to the situation where the machine learning model is not learning enough of the signal from the input data and leads to poor performance in terms of final predictions. In such a case, the model is too simple to approximate the output based on the given inputs. On the other hand, high variance refers to overfitting (learning too much on training data). In the case of high variance, the learning of the model on the training data affects the generalization performance on the unseen or test data due to an overcomplex model. One needs to balance the bias versus variance as both are opposite of each other. In other words, if we increase bias, the variance goes down, and vice versa, as shown in Figure 1-8.

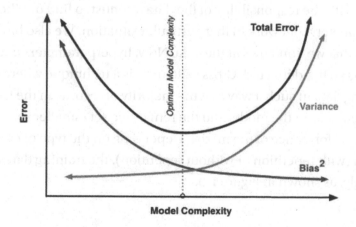

Figure 1-8. *Bias versus variance*

Cross Validation and Hyperparameters

For most of the machine learning algorithms out there, there is a set of hyperparameters that can be adjusted accordingly to have the best performance coming out of the model. The famous analogy of the hyperparameters is that of tuning knobs in a radio/transistor to match the exact frequency of the radio station to hear the sound properly. Likewise, hyperparameters provide the best possible combination for a model's performance for a given training data. The following are a few examples of hyperparameters in the case of a machine learning model such as random forest:

- Number of trees

- Maximum number of features

- Maximum depth of trees

For the different values of the previous hyperparameters, the model would learn the different parameters for the given input data, and the prediction performance would vary accordingly. Most libraries provide the default value of these parameters for the vanilla version of the model, and it's the responsibility of the data scientist to find out the best hyperparameters that work in that particular situation. We also have to be careful that we don't overfit the data. Now, hyperparameters and cross validations go hand in hand. Cross validation is a technique where we split the training data in such a way that the majority of records in the training set are used to train the model and the remaining set (smaller set) is used to test the performance of the model. Depending on the type of cross validation (with repetition or without repetition), the training data is split accordingly, as shown in Figure 1-9.

Figure 1-9. *Cross validation*

Performance Metrics

There are different ways in which the performance of a machine learning model can be evaluated depending on the nature of algorithm used. As mentioned previously, there are broadly two categories of models: regression and classification. For the models that predict a continuous target, such as R-square, root mean squared error (RMSE) can be used, whereas for the latter, an accuracy measure is the standard metric. However, the cases where there is class imbalance and the business needs to focus on only one out of the positive or negative class, measures such as precision and recall can be used.

Now that we have gone over the fundamentals and important concepts in machine learning, it's time for us to build a simple machine learning model on a cloud platform, namely, Databricks.

Databricks is an easy and convenient way to get started with cloud infrastructure to build and run machine learning models (single-threaded as well as distributed). I have given a deep introduction of the Databricks platform in a couple of my earlier books (*Machine Learning Using PySpark* and *Learn PySpark*). The objective of this section in this chapter is to give you a flavor of how to get up and running with ML on the cloud by just signing up for any of the major cloud services providers (Google, Amazon, Microsoft, Databricks). Most of these platforms allows users to simply sign up and use the ML services (in some cases with limited capabilities) for a predefined period or up to the extent of exhausting the free credit points. Databricks allows you to use the community edition of its platform that offers up to 6 GB of cluster size. We are going to use the community edition to build and understand a decision tree model on a fake currency dataset. The dataset contains four attributes of the currency notes that can be used to detect whether a currency note is genuine or fake. Since we are using the community edition, there is a limitation on the size of the dataset, and hence it's been kept relatively small for demo purpose.

Note   Sign up for the Databricks community edition to run this code.

The first step is to start a new cluster with the default settings as we are not building a complicated model here. Once the cluster is up and running, we need to simply upload the data to Databricks from the local system. The next step is to create a new notebook and attach it to the cluster we created earlier. The next step is to import all required libraries and confirm that the data was uploaded successfully.

```
[In]: import pandas as pd
[In]: import numpy as np
[In]: from sklearn.model_selection import train_test_split
[In]: from sklearn.tree import DecisionTreeClassifier
[In]: from sklearn.metrics import classification_report
```

The following line of command will show the table (dataset) that was uploaded from the local system:

```
[In]: display(dbutils.fs.ls("/FileStore/tables/"))
```

The next step is to create a Spark dataframe from the table and later convert it to a pandas dataframe to build the model.

```
[In:sparkDF=spark.read.csv('/FileStore/tables/currency_note_
data.csv', header="true", inferSchema="true")
```

```
[In]: df=sparkDF.toPandas()
```

We can take a look at the top five rows of the dataframe by using the pandas <u>head</u> function. This confirms that we have a total of five columns including the target column (<u>Class</u>).

```
[In]: df.head(5)
```

```
[Out]:
```

	Variance	Skewness	Curtosis	Entropy	Class
0	3.62160	8.6661	-2.8073	-0.44699	0
1	4.54590	8.1674	-2.4586	-1.46210	0
2	3.86600	-2.6383	1.9242	0.10645	0
3	3.45660	9.5228	-4.0112	-3.59440	0
4	0.32924	-4.4552	4.5718	-0.98880	0

As mentioned earlier, the data size is relatively small, and we can see that it contains just 1,372 records in total, but the target class seems to be well balanced, and hence we are not dealing with an imbalanced class.

```
[In]: df.shape
[Out]: (1372, 5)
```

```
[In]: df.Class.value_counts()
```

```
[Out]:
0 762
1 610
```

We can also check whether there are any missing values in the dataframe by using the info function. The dataframe seems to contain no missing values as such.

```
[In]: df.info()
```

```
[Out]:
```

```
<class 'pandas.core.frame.DataFrame'>
RangeIndex: 1372 entries, 0 to 1371
Data columns (total 5 columns):
Variance    1372 non-null float64
Skewness    1372 non-null float64
Curtosis    1372 non-null float64
Entropy     1372 non-null float64
Class       1372 non-null int32
dtypes: float64(4), int32(1)
memory usage: 48.3 KB
```

The next step is to split the data into training and test sets using the train test split functionality

```
[In]: X = df.drop('Class', axis=1)
[In]: y = df['Class']
```

```
[In]:X_train,X_test,y_train,y_test=train_test_split(X,y,test_
    size=0.25,random_state=30)
```

Now that we have the training set separated out, we can build a decision tree with default hyperparameters to keep things simple. Remember, the objective of building this model is simply to introduce the process of training a model on a cloud platform. If you want to train a much more complicated model, please feel free to add your own steps such as enhanced feature engineering, hyperparameter tuning, baseline models, visualization, or more. We are going to build much more complicated models that include all the previous steps in later chapters of this book.

```
[In]: dec_tree=DecisionTreeClassifier().fit(X_train,y_train)
```

```
[In]: dec_tree.score(X_test,y_test)
```

```
[Out]: 0.9854227405247813
```

We can see that the decision tree seems to be doing incredibly well on the test data. We can also go over the other performance metrics apart from accuracy using the classification report function.

```
[In]: y_preds = dec_tree.predict(X_test)
[In]: print(classification_report(y_test,y_preds))
```

```
[Out]:
```

	precision	recall	f1-score	support
0	0.99	0.98	0.99	178
1	0.98	0.99	0.98	165
micro avg	0.99	0.99	0.99	343
macro avg	0.99	0.99	0.99	343
weighted avg	0.99	0.99	0.99	343

Deep Learning

In this section of the chapter, we will go over the fundamentals of deep learning and its underlying operating principles. Deep learning has been in the limelight for quite a few years now and is improving leaps and bounds in terms of solving various business challenges. From image captioning to language translation to self-driving cars, deep learning has become an important component in the larger scheme of things. To give you an example, Google's products such as Gmail, YouTube, Search, Maps, and Assistance are all using deep learning in some or the other way in the background due to its incredible ability to provide far better results compared to some of the other traditional machine learning algorithms.

But what exactly is deep learning? Well, before even getting into deep learning, we must understand what neural networks are. Deep learning in fact is sort of an extension to the neural network. As mentioned earlier in the chapter, neural networks are not new, but they didn't take off due to various limitations. Those limitations don't exist anymore, and businesses and research community are able to leverage the true power of neural networks now.

In supervised learning settings, there is a specific input and corresponding output. The objective of the machine learning algorithms is to use this data and approximate the relationship between input and output variables. In some cases, this relationship is evident and easy to capture, but in realistic scenarios, the relationship between the input and output variables is complex and nonlinear in nature. To give an example, for a self-driving car, the input variables could be as follows:

- Terrain

- Distance from nearest object

- Traffic light

- Sign boards

The output needs to be either turn, drive fast or slowly, apply brakes, etc. As you might think, the relationship between input variables and output variables is pretty complex in nature. Hence, the traditional machine learning algorithm finds it hard to map this kind of relationship. Deep learning outperforms machine learning algorithms in such situations as it is able to learn those nonlinear features as well.

Human Brain Neuron vs. Artificial Neuron

As mentioned, deep learning is extension of neural networks only and also known as *deep neural networks*. Neural networks are a little different compared to other machine learning algorithms in terms of learning. Neural networks are loosely inspired by neurons in the human brain. Neural networks are made up of artificial neurons. Although I don't claim to be an expert of neuroscience or functioning of the brain, let me try to give you a high-level overview of "how the human brain functions." As you might be already aware, the human brain is made up of billions of neurons and an incredible number of connections between them. Each neuron is connected to multiple other neurons, and they repeatedly exchange information (signal). Each activity that we do physically or mentally fires up a certain set of neurons in our brains. Now, every single neuron consists of three basic components.

- Dendrites

- Cell body

- Terminals

As we can see in Figure 1-10, the dendrites are responsible for receiving the signal from other neurons. A dendrite act as a receiver to the particular neuron and passes information to the cell body where this specific information is processed. Now, based on the level of information, it either activates (fires up) or doesn't trigger. This activity depends on a particular threshold value of the neuron. If the incoming signal value is below that threshold, it would not fire; otherwise, it activates. Finally, the third component are the terminals that are connected with dendrites of other neurons. Terminals are responsible for passing on the output of the particular neuron to other relevant connections.

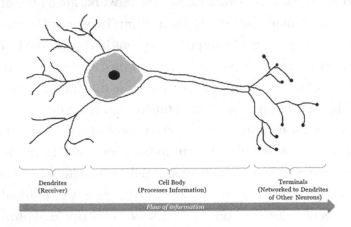

Figure 1-10. *Neuron*

Now, we come to the artificial neuron, which is the basic building block of a neural network. A single artificial neuron consists of two parts mainly; one is the summation, and other is activation, as shown in Figure 1-11. This is also known as a *perceptron*. Summation refers to adding all the input signals, and activation refers to deciding whether the neuron would trigger or not based on the threshold value.

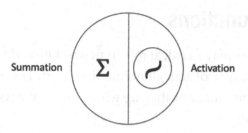

Figure 1-11. *Artificial neuron*

Let's say we have two binary inputs (X1, X2) and the weights of their respective connections (W1, W2). The weights can be considered similar to the coefficients of input variables in traditional machine learning. These weights indicate how important the particular input feature is in the model. The summation function calculates the total sum of the input. The activation function then uses this total summated value and gives a certain output, as shown in Figure 1-12. Activation is sort of a decision-making function. Based on the type of activation function used, it gives an output accordingly. There are different types of activation functions that can be used in a neural network layer.

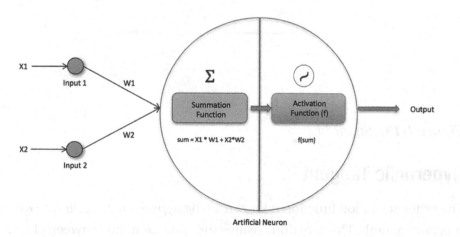

Figure 1-12. *Neuron calculation*

Activation Functions

Activation functions play a critical role in neural networks as the output varies based on the type of activation function used. There are typically four main activation functions that are widely used. We will briefly cover these in this section.

Sigmoid Activation Function

The first type of activation function is a sigmoid function. This activation function ensures the output is always between 0 and 1 irrespective of the input, as shown in Figure 1-13. That's why it is also used in logistic regression to predict the probability of the event.

$$f(x) = \frac{1}{1 + e^{-x}}$$

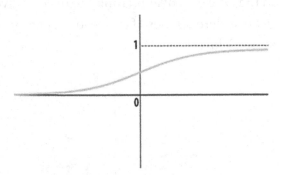

Figure 1-13. *Sigmoid*

Hyperbolic Tangent

The other activation function is known as the *hyperbolic tangent activation function*, or tanh. This function ensures the value remains between -1 to 1 irrespective of the output, as shown in Figure 1-14. The formula of the tanh activation function is as follows:

$$f(x) = \frac{e^{2x} - 1}{e^{2x} + 1}$$

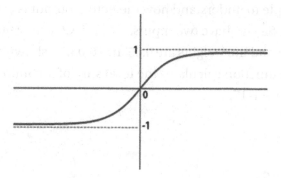

Figure 1-14. *Tanh*

Rectified Linear Unit

The rectified linear unit (relu) has been really successful over the past couple of years and has become the default choice for the activation function. It is powerful as it produces a value between 0 and ∞. If the input is 0 or less than 0, then the output is always going to be 0, but for anything more than 0, the output is similar to the input, as shown in Figure 1-15. The formula for relu is as follows:

$$f(x) = \max(0, x)$$

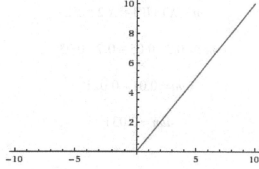

Figure 1-15. *Relu*

Neuron Computation Example

Since we have basic understanding of different activation functions, let's look at an example to understand how the actual output is calculated inside a neuron. Say we have two inputs, X1 and X2, with values of 0.2 and 0.7, respectively, and the weights are 0.05 and 0.03, as shown in Figure 1-16. The summation function calculates the total sum of incoming input signals as shown in Figure 1-17.

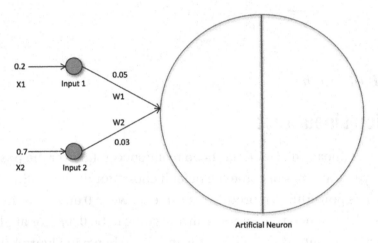

Figure 1-16. *Neuron input*

Here is the summation:

$$sum = X1 * W1 + X2 * W2$$

$$sum = 0.2 * 0.05 + 0.7 * 0.03$$

$$sum = 0.01 + 0.021$$

$$sum = 0.031$$

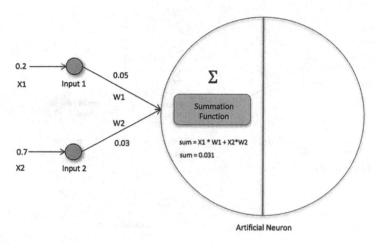

Artificial Neuron

Figure 1-17. *Summation*

The next step is to pass this sum through an activation function. Let's consider using a sigmoid function that returns values between 0 and 1 irrespective of the input. The sigmoid function would calculate the value as shown here and in Figure 1-18:

$$f(x) = \frac{1}{(1 + e^{-x})}$$

$$f(sum) = \frac{1}{(1 + e^{-sum})}$$

$$f(0.031) = \frac{1}{(1 + e^{-0.031})}$$

$$f(0.031) = 0.5077$$

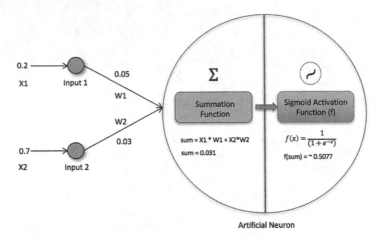

Figure 1-18. *Activation*

So, the output of this single neuron is equal to 0.5077.

Neural Network

When we combine multiple neurons, we end up with a neural network. The simplest and most basic neural network can be built using just the input and output neurons, as shown in Figure 1-19.

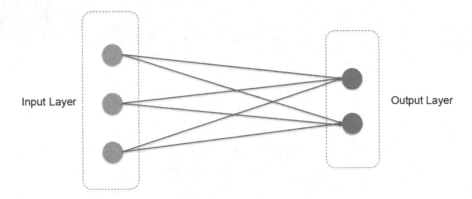

Figure 1-19. *Simple network*

The challenge with using a neural network like this is that it can only learn linear relationships and cannot perform well in cases where the relationship between the input and output is nonlinear. As we have already seen, in real-world scenarios, the relationship is hardly simple and linear. Hence, we need to introduce an additional layer of neurons between the input and output layers to increase its capability to learn different kinds of nonlinear relationships as well. This additional layer of neurons is known as the *hidden layer*, as shown in Figure 1-20. It is responsible for introducing the nonlinearities into the learning process of the network. Neural networks are also known as *universal approximators* since they carry the ability to approximate any relationship between the input and output variables no matter how complex and nonlinear it is nature. A lot depends on the number of hidden layers in the networks and the total number of neurons in each hidden layer. Given enough hidden layers, it can perform incredibly well at mapping this relationship.

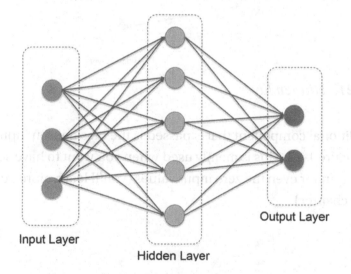

Figure 1-20. *Neural network with hidden layer*

Training Process

A neural network is all about the various connections (red lines) and different weights associated with these connections. The training of neural networks primarily includes adjusting these weights in such a way that the model can predict with a higher amount of accuracy. To understand how neural networks are trained, let's break down the steps of network training.

Step 1: Take the input values as shown in Figure 1-21 and calculate the output values that are passed to hidden neurons. The weights used for the first iteration of the sum calculation are generated randomly.

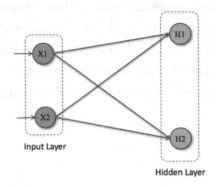

Figure 1-21. *Hidden layer*

An additional component that is passed is the bias neuron input, as shown in Figure 1-22. This is mainly used when you want to have some nonzero output for even the zero input values (you'll learn more about bias later in the chapter).

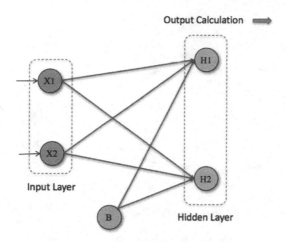

Figure 1-22. *Bias component*

Step 2: The network-predicted output is compared with the actual output, as shown in Figure 1-23.

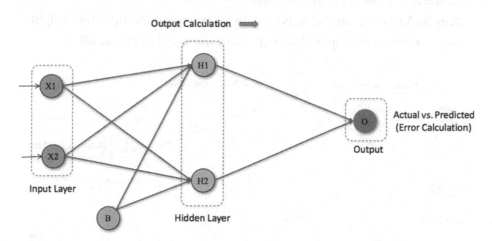

Figure 1-23. *Output comparison*

Step 3: The error is back propagated to the network, as shown in Figure 1-24.

33

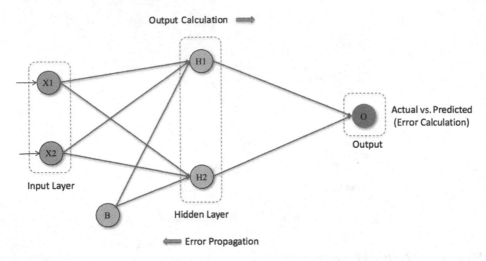

Figure 1-24. *Error propogation*

Step 4: The weights are re-adjusted according to the output to minimize the errors, as shown in Figure 1-25.

Step 5: A new output value is calculated based on the updated weights. Step 2 repeats until no more changes in the weights are possible.

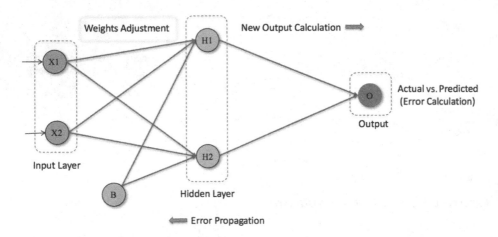

Figure 1-25. *Weight adjustment*

Role of Bias in Neural Networks

One common question that people have is, why do we add bias in neural networks? Well, the role of bias is critical for the right learning of the model as it a direct relation to the performance of the model. To understand the role of bias in neural networks, we need to go back to linear regression and uncover the role of intercepts in the regression line. We know for a fact that the value of intercept changes the position of a line to up or down, whereas slope changes the angle of the line, as shown in Figure 1-26. If the slope is less than the input, the variable has less impact on the final prediction because for changes in value in the input, the corresponding output change is less for a small slope, whereas if the slope value if higher, then the output is more sensitive toward the smallest change in the input value, as shown in Figure 1-26.

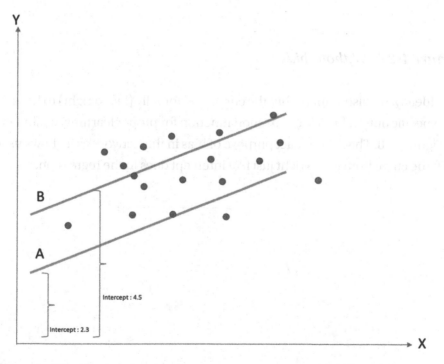

Figure 1-26. *Slope in regression*

Hence, the slope value decides the angle at which the line exists, whereas the intercept value decides at what position the line exists (low or high).

In the same way, if we don't use bias for a network, the simple calculation would be the combined output from the weight and the input to the activation function. Since the inputs are fixed and only the weights can be altered, we can only change the steepness/angle of the activation curve, which is only half the work done (although some cases it works), as shown in Figure 1-27.

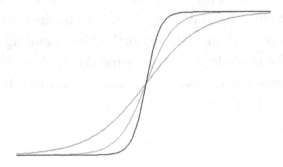

Figure 1-27. *Without bias*

Ideally, we also want to shift the curve horizontally (left to right) to have the specific output from the activation function for proper learning, as shown in Figure 1-28. That is the exact purpose of bias in the network as it allows us to shift the curve from left to right just like intercept does in the regression.

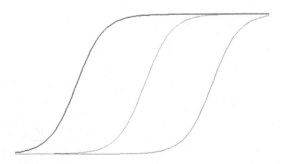

Figure 1-28. *With bias*

Now that we have a good understanding of how deep learning works, we can dive into specific neural networks that are widely used. There are many variants of deep learning models, but we are going to focus on two types of deep learning models only.

- Convolutional neural networks (CNNs)

- Recurrent neural networks (RNNs)

CNN

There was a major breakthrough when CNNs were first used for image-based tasks. The amount of accuracy they provided surpassed every other algorithm that was previously used for image recognition. Since then, there are multiple variants of CNNs being used with added capabilities to solve specific image-based tasks such as face recognition, computer vision, self-driving cars, etc. The CNNs have the ability to extract high-level features from the images that capture the most important aspects of the image for recognition via the process known as *convolution*, as shown in Figure 1-29.

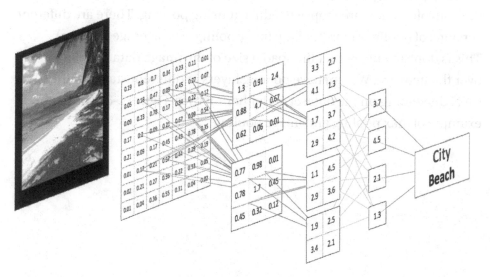

Figure 1-29. *Image classification*

Convolution is an easy process to understand as we roll a filter (also called as *kernel*) over the image pixel values to extract the convoluted feature, as shown in Figure 1-30.

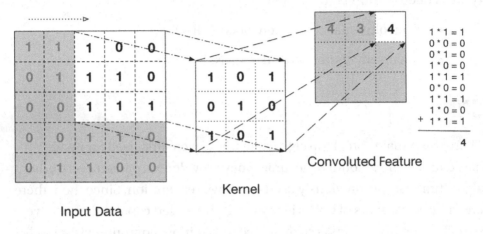

Figure 1-30. *Image convolution*

Rolling of the filter over the image indicates that we take a dot product of the specific region of values of the image with the filter. Once we have the convoluted feature map, we reduce it using pooling. There are different versions of pooling (max pooling, min pooling, and average pooling). This is done to ensure that the spatial size of the image data is reduced over the network. We can have pooling layers at different stages of the CNN depending on the dataset and other metrics. Figure 1-31 shows the example of max pooling and image pooling.

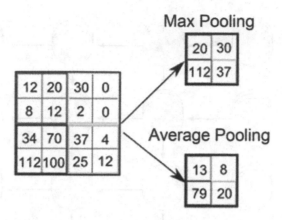

Figure 1-31. Pooling

We can repeat the previous steps (convolution and pooling) multiple times in a network to learn the main features of the image and finally pass it to the fully connected layer at the end to make the classification.

RNN

The general feedforward neural networks and CNNs are not good for time-series kinds of datasets as these networks don't have any memory of their own. Recurrent neural networks bring with them the unique ability to remember important stuff during the training over a period of time. This makes them well suited for tasks such as natural language translation, speech recognition, and image captioning. These networks have states defined over a timeline and use the output of the previous state in the current input, as shown in Figure 1-32.

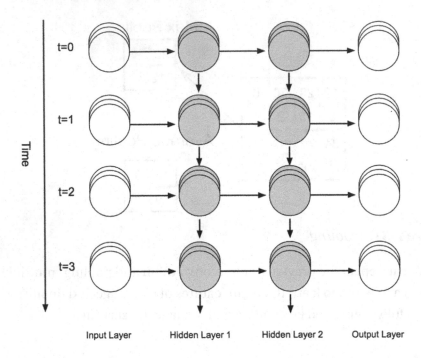

Figure 1-32. *RNN*

Although RNNs have proved to be really effective in time-series kinds of applications, it does run into some serious limitations in terms of performance because of its architecture. It struggles with what is known as a *vanishing gradient problem* that occurs due to no or very little updates in the weights of the network as the network tries to use data points that are at the early stages of timeline. Hence, it has a limited memory to put it in simple terms. To tackle this problem, there are couple of other variants of RNNs.

- Long short-term memory (LSTM)

- Gradient recurring unit (GRU)

- Attention networks (encoder-decoder model)

Now we are going to use a small dataset and build a deep learning model to predict the sentiment given the user review. We are going to make use of TensorFlow and Keras to build this model. There are couple of steps that we need to do before we train this model in Databricks. We first need to go to the cluster and click Libraries. On the Libraries tab, we need to select the Pypi option and mention Keras to get it installed. Similarly, we need to mention TensorFlow as well once Keras is installed.

Once we upload the reviews dataset, we can create a pandas dataframe like we did in the earlier case.

```
[In]: from tensorflow.keras.models import Sequential
[In]: from tensorflow.keras.layers import LSTM,Embedding
[In]: from tensorflow.keras.layers import Dense
[In]: from tensorflow.keras.preprocessing.text import Tokenizer
[In]: from tensorflow.keras.preprocessing.sequence import pad_
      sequences
[In]:sparkDF= spark.read.csv('/FileStore/tables/text_summary.
     csv', header="true", inferSchema="true")

[In]: df=sparkDF.toPandas()

[In]: df.columns
[Out]: Index(['Sentiment', 'Summary'], dtype='object')
```

As we can see, there are just two columns in the dataframe.

```
[In]: df.head(10)

[Out]:
```

	Sentiment	Summary
0	1	Yummy Flavor
1	0	tastes like soap
2	0	Limited applications
3	1	tasty in the keurig
4	1	Good snack food
5	1	Unusual, delicate yet flavorful. I began deve...
6	1	So delicious!!
7	1	5 Hour Energy drink
8	1	Great baby food - buy somewhere else
9	1	Instinct Food Saved My Cats Life

```
[In]: df.Sentiment.value_counts()
```

```
[Out]:
1 1000
0 1000
```

We can also confirm the class balance by taking a value counts of the target column. It seems the data is well balanced. Before we go ahead with building the model, since we are dealing with text data, we need to clean it a little bit to ensure no unwanted errors are thrown at the time of training. Hence, we write a small helper function using regular expressions.

```
[In]:
import re
def clean_reviews(text):
    text=re.sub("[^a-zA-Z]"," ",str(text))
    return re.sub("^\d+\s|\s\d+\s|\s\d+$", " ", text)
```

```
[In]: df['Summary']=df.Summary.apply(clean_reviews)
```

```
[In]: df.head(10)
```

[Out]:

	Sentiment	Summary
0	1	Yummy Flavor
1	0	tastes like soap
2	0	Limited applications
3	1	tasty in the keurig
4	1	Good snack food
5	1	Unusual delicate yet flavorful I began deve...
6	1	So delicious
7	1	Hour Energy drink
8	1	Great baby food buy somewhere else
9	1	Instinct Food Saved My Cats Life

The next step is to separate input and output data. Since the data is already small, we are not going to split it into train and test sets; rather, we will train the model on all the data.

```
[In]: X=df.Summary
[In]: y=df.Sentiment
```

We now create the tokenizer object with 10,000 vocab words, and an out-of-vocabulary (oov) token is mentioned for the unseen words that the model gets exposed to that are not part of the training.

```
[In]: tokenizer=Tokenizer(num_words=10000,oov_token='xxxxxxx')
```

```
[In]: tokenizer.fit_on_texts(X)
```

```
[In]: X_dict=tokenizer.word_index
```

```
[In]: len(X_dict)
```

```
[Out]: 2018
```

```
[In]: X_dict.items()
```

```
[Out]:
```

```
Out[26]: dict_items([('xxxxxx', 1), ('not', 2), ('great', 3), ('the', 4), ('good', 5), ('a', 6), ('for', 7), ('it', 8), ('and', 9), ('i', 10), ('this', 11), ('my', 12), ('to', 13), ('is', 14), ('o
f', 15), ('but', 16), ('coffee', 17), ('taste', 18), ('like', 19), ('in', 20), ('best', 21), ('love', 22), ('t', 23), ('product', 24), ('tea', 25), ('very', 26), ('flavor', 27), ('as', 28), ('food',
29), ('too', 30), ('s', 31), ('dog', 32), ('bad', 33), ('these', 34), ('no', 35), ('so', 36), ('price', 37), ('made', 38), ('you', 39), ('tasty', 40), ('what', 41), ('delicious', 42), ('with', 43),
('just', 44), ('tastes', 45), ('excellent', 46), ('don', 47), ('chocolate', 48), ('me', 49), ('much', 50), ('that', 51), ('ever', 52), ('tasting', 53), ('one', 54), ('loves', 55), ('free', 56), ('yu
mmy', 57), ('are', 58), ('sweet', 59), ('horrible', 60), ('at', 61), ('stuff', 62), ('if', 63), ('have', 64), ('they', 65), ('all', 66), ('dogs', 67), ('them', 68), ('k', 69), ('cup', 70), ('more',
71), ('buy', 72), ('on', 73), ('little', 74), ('your', 75), ('treat', 76), ('ok', 77), ('healthy', 78), ('cookies', 79), ('stale', 80), ('nice', 81), ('than', 82), ('awful', 83), ('sugar', 84), ('re
ally', 85), ('be', 86), ('chips', 87), ('cat', 88), ('terrible', 89), ('was', 90), ('gluten', 91), ('cats', 92), ('worth', 93), ('from', 94), ('con', 95), ('favorite', 96), ('by', 97), ('china', 9
8), ('money', 99), ('wonderful', 100), ('quality', 101), ('better', 102), ('do', 103), ('up', 104), ('beware', 105), ('poor', 106), ('worst', 107), ('disappointed', 108), ('pretty', 109), ('an', 11
0), ('candy', 111), ('nothing', 112), ('awesome', 113), ('weak', 114), ('works', 115), ('only', 116), ('strong', 117), ('oil', 118), ('organic', 119), ('water', 120), ('packaging', 121), ('out', 12
2), ('fresh', 123), ('yum', 124), ('mix', 125), ('first', 126), ('perfect', 127), ('snack', 128), ('hot', 129), ('there', 130), ('amazing', 131), ('chicken', 132), ('value', 133), ('butter', 134),
('disappointing', 135), ('way', 136), ('salty', 137), ('treats', 138), ('amazon', 139), ('will', 140), ('did', 141), ('cups', 142), ('work', 143), ('waste', 144), ('again', 145), ('get', 146), ('har
d', 147), ('wow', 148), ('popcorn', 149), ('has', 150), ('got', 151), ('well', 152), ('eat', 153), ('old', 154), ('natural', 155), ('peanut', 156), ('sick', 157), ('fantastic', 158), ('easy', 159),
('breakfast', 160), ('order', 161), ('deal', 162), ('had', 163), ('dry', 164), ('ve', 165), ('bought', 166), ('new', 167), ('about', 168), ('didn', 169), ('warning', 170), ('find', 171), ('green', 1
72), ('off', 173), ('small', 174), ('expensive', 175), ('were', 176), ('does', 177), ('canned', 178), ('yuck', 179), ('blend', 180), ('thing', 181), ('doesn', 182), ('down', 183), ('must', 184), ('s
mell', 185), ('make', 186), ('protein', 187), ('batch', 188), ('jerky', 189), ('ingredients', 190), ('even', 191), ('coconut', 192), ('drink', 193), ('baby', 194), ('light', 195), ('something', 19
6), ('full', 197), ('okay', 198), ('tiny', 199), ('go', 200), ('espresso', 201), ('most', 202), ('roast', 202), ('loved', 204), ('try', 205), ('right', 206), ('would', 207), ('beans', 208), ('bitte
r', 209), ('tasteless', 210), ('think', 211), ('overpriced', 212), ('never', 213), ('use', 214), ('or', 215), ('else', 216), ('tasted', 217), ('spicy', 218), ('mild', 219), ('contains', 220), ('can
```

As we can see, there are 2,018 unique words in the training data. Now we transform each review into a numerical vector based on the token mapping done using tokenizer.

```
[In]: X_seq=tokenizer.texts_to_sequences(X)
```

```
[In]: X_seq[:10]
```

```
[Out]:
```

```
Out[28]: [[57, 27],
 [45, 19, 547],
 [548, 855],
 [40, 20, 4, 255],
 [5, 128, 29],
 [856, 549, 321, 550, 10, 857, 858, 859, 7, 8],
 [36, 42],
 [551, 256, 193],
 [3, 194, 29, 72, 860, 216],
 [861, 29, 552, 12, 92, 322]]
```

Although the text-to-sequence function converted each review into a vector, there is a slight problem as each vector's length is different based on the length of the original review. To fix this issue, we make use of the padding function. It ensures that each vector is conformed to a fix length (we add a set of 0s at the end or beginning depending on the type of padding used: pre or post).

```
[In]: X_padded_seq=pad_sequences(X_
seq,padding='post',maxlen=100)
```

```
[In]: X_padded_seq[:3]
```

```
[Out]:
```

```
Out[30]: array([[ 57,  27,   0,   0,   0,   0,   0,   0,   0,   0,   0,   0,   0,
           0,   0,   0,   0,   0,   0,   0,   0,   0,   0,   0,   0,
           0,   0,   0,   0,   0,   0,   0,   0,   0,   0,   0,   0,   0,
           0,   0,   0,   0,   0,   0,   0,   0,   0,   0,   0,   0,   0,
           0,   0,   0,   0,   0,   0,   0,   0,   0,   0,   0,   0,   0,
           0,   0,   0,   0,   0,   0,   0,   0,   0,   0,   0,   0,   0,
           0,   0,   0,   0,   0,   0,   0,   0,   0,   0,   0,   0,   0,
           0,   0,   0,   0,   0,   0,   0,   0,   0],
         [ 45,  19, 547,   0,   0,   0,   0,   0,   0,   0,   0,   0,   0,
           0,   0,   0,   0,   0,   0,   0,   0,   0,   0,   0,   0,   0,
           0,   0,   0,   0,   0,   0,   0,   0,   0,   0,   0,   0,   0,
           0,   0,   0,   0,   0,   0,   0,   0,   0,   0,   0,   0,   0,
           0,   0,   0,   0,   0,   0,   0,   0,   0,   0,   0,   0,   0,
           0,   0,   0,   0,   0,   0,   0,   0,   0,   0,   0,   0,   0,
           0,   0,   0,   0,   0,   0,   0,   0,   0,   0,   0,   0,   0,
           0,   0,   0,   0,   0,   0,   0,   0,   0],
         [548, 855,   0,   0,   0,   0,   0,   0,   0,   0,   0,   0,   0,
           0,   0,   0,   0,   0,   0,   0,   0,   0,   0,   0,   0,   0,
           0,   0,   0,   0,   0,   0,   0,   0,   0,   0,   0,   0,   0,
           0,   0,   0,   0,   0,   0,   0,   0,   0,   0,   0,   0,   0,
           0,   0,   0,   0,   0,   0,   0,   0,   0,   0,   0,   0,   0,
           0,   0,   0,   0,   0,   0,   0,   0,   0,   0,   0,   0,   0,
           0,   0,   0,   0,   0,   0,   0,   0,   0,   0,   0,   0,   0,
           0,   0,   0,   0,   0,   0,   0,   0,   0]], dtype=int32)
```

```
[In]: X_padded_seq.shape
```

```
[Out]: (2000, 100)
```

As we can see, we have now every review that has been converted into a fixed-size vector. In the next step, we flatten our target variable and declare some of the global parameters for the network. You can choose your own parameter values.

```
[In]: y = np.array(y)
[In]: y=y.flatten()

[In]: max_length = 100
[In]: vocab_size = 10000
[In]: embedding_dims = 50
```

Now we build the model that is sequential in nature and makes use of the relu activation function.

```
[In]: model = tf.keras.Sequential([
tf.keras.layers.Embedding(input_length=100,input_
dim=10000,output_dim=50),
tf.keras.layers.Flatten(),
tf.keras.layers.Dense(50, activation='relu'),
tf.keras.layers.Dense(1, activation='sigmoid')
])
[In]:model.compile(loss='binary_crossentropy',optimizer='adam',
metrics=['accuracy'])
[In]: model.summary()
```

[Out]:

```
Model: "sequential_1"

_____
Layer (type)                 Output Shape              Param #
=================================================================
embedding_1 (Embedding)      (None, 100, 50)           500000
_____
flatten_1 (Flatten)          (None, 5000)              0
_____
dense_2 (Dense)              (None, 50)                250050
_____
dense_3 (Dense)              (None, 1)                 51
=================================================================
Total params: 750,101
Trainable params: 750,101
Non-trainable params: 0
```

```
[In]: num_epochs = 10
[In]: model.fit(X_padded_seq,y, epochs=num_epochs)
```

The model seems to be learning really well, but there are chances of overfitting the data as well. We will deal with overfitting and other settings of a network in later chapters of the book.

```
Train on 2000 samples
Epoch 1/10

  32/2000 [..............................] - ETA: 0s - loss: 8.1873e-04 - accuracy: 1.0000
 320/2000 [===>..........................] - ETA: 0s - loss: 0.0021 - accuracy: 1.0000
 576/2000 [=======>......................] - ETA: 0s - loss: 0.0135 - accuracy: 0.9931
 832/2000 [===========>..................] - ETA: 0s - loss: 0.0098 - accuracy: 0.9952
1088/2000 [================>.............] - ETA: 0s - loss: 0.0101 - accuracy: 0.9954
1344/2000 [===================>..........] - ETA: 0s - loss: 0.0109 - accuracy: 0.9955
1600/2000 [=======================>......] - ETA: 0s - loss: 0.0116 - accuracy: 0.9944
1856/2000 [==========================>...] - ETA: 0s - loss: 0.0126 - accuracy: 0.9935
2000/2000 [==============================] - 0s 214us/sample - loss: 0.0136 - accuracy: 0.9930
```

```
Epoch 10/10

  32/2000 [..............................] - ETA: 0s - loss: 5.4804e-04 - accuracy: 1.0000
 288/2000 [===>..........................] - ETA: 0s - loss: 0.0084 - accuracy: 0.9965
 512/2000 [======>.......................] - ETA: 0s - loss: 0.0126 - accuracy: 0.9922
 800/2000 [===========>..................] - ETA: 0s - loss: 0.0116 - accuracy: 0.9925
1056/2000 [==============>...............] - ETA: 0s - loss: 0.0151 - accuracy: 0.9905
1312/2000 [=================>............] - ETA: 0s - loss: 0.0157 - accuracy: 0.9901
1568/2000 [=====================>........] - ETA: 0s - loss: 0.0153 - accuracy: 0.9904
1824/2000 [==========================>...] - ETA: 0s - loss: 0.0148 - accuracy: 0.9907
2000/2000 [=============================] - 0s 208us/sample - loss: 0.0136 - accuracy: 0.9915
```

Now that we have gotten some exposure to both machine learning and deep learning fundamentals and a sense of how to build models in cloud, we can look at different applications of ML/DL in businesses around the world along with some of the challenges that come with it.

Industrial Applications and Challenges

In the final section of this chapter, we will go though some of the real applications of ML and AI. Businesses are heavily investing in ML and AL across the globe and establishing standard procedures to leverage the capabilities of ML and AI to build their competitive edge. There are multiple areas where ML and AI are being currently applied and providing great value to businesses. We will look at few of the major domains where ML and AI are transforming the landscape.

Retail

One of the business verticals that is making incredible use of ML and AI is retail. Since retail business generates a lot of customer data, it offers a perfect platform for applying ML and AI. The retail sector has always faced multiple challenges such as out-of-stock situations, suboptimal pricing, limited cross sell or upsell, and inadequate personalization. ML and AI have been able attack many of these challenges and offer incredible impact in retail space. There are numerous applications that have been built in the retail space that are powered by ML and AL in the last decade, and the number continues to grow. The most prominent application is the recommender system. Online retail businesses are thriving on recommender systems as they can increase their revenue by a great deal. In addition, retail uses ML and AI capabilities for stock optimization to control the inventory levels and reduce costs. Dynamic pricing is another area where AI and ML are being used comprehensively to get maximum returns. Customer segmentation is also done using ML as it uses not only the demographics information of the customer but the transactional data and takes multiple other variables into consideration before revealing the different groups within the customer base. Product categorization is also being done using ML as it saves a huge amount of manual effort and increases the accuracy levels of labeling the products. Demand forecasting and stock optimization are tackled using ML and AI to save costs. Route planning has also been handled by ML and AI in the last few years as it enables businesses to fulfill orders in more effective way. As a result of ML and AI applications in retail, the cost savings have improved, businesses are able to take informed decisions, and the overall customer satisfaction has gone up.

Healthcare

Another business vertical to be deeply impacted by ML and AI is healthcare. Diagnoses based on image data using ML and AI are being adopted at a quick rate across healthcare spectrum. The prime reasons are the levels of accuracy levels offered by ML and AI and the ability to learn from data of past decades. ML and AI algorithms on X-rays, MRI scans, and various other images in the healthcare domain are being heavily used to detect any anomalies. Virtual assistants and chatbots are also being deployed as part of applications to assist with explaining lab reports. Finally, insurance verification is also being done using ML models in healthcare to avoid any inconsistency.

Finance

The finance domain has always had data, lots of it. Out of any other domain, finance has always been data enriched. Hence, there are multiple applications being built over the last decade based on ML and AI. The most prominent one is the fraud detection system, which used anomaly detection algorithms in the background. Other areas are portfolio management and algorithmic trading. ML and AI have the ability to scan more than 100 years of past data and learn the hidden patterns to suggest the best calibration of a portfolio. Complex AI systems are being used to make extremely fast decisions about trading to maximize the gains. ML and AI are also used in risk mitigation and loan insurance underwriting. Again, recommender systems are being used to upsell and cross sell various financial products by various institutions. They also use recommender systems to predict the churn of the customer base in order to formulate a strategy to retain the customers who are likely to discontinue with a specific product or service. Another important usage of ML and AI in the finance sector is to check whether the loan should be granted or not to various applicants based on predictions made by the model. In addition, ML is being used to validate whether the insurance claims are genuine or fraud based on the ML model predictions.

Travel and Hospitality

Just like retail, the travel and hospitality domain is thriving on ML- and AI-based applications. To name a few, recommender systems, price forecasting, and virtual assistants are all ML- and AI-based applications that are being leveraged in the travel and hospitality vertical. From recommending best deals to alternative travel dates, recommender systems are super-critical to drive customer behavior in this sector. It also recommends new travel destinations based on a user's preferences, which are highly tailored using ML in the background. AI is also being used to sending timely alerts to customers by predicting future price movements based on the various factors. Virtual assistants nowadays are part of every travel website as the customers don't want to wait to get the relevant information. On top of that, the interactions with these virtual assistants are very human like as natural language intelligence is already being embedded into these chat bots to a great extent so as to understand simple questions and reply in a similar manner.

Media and Marketing

Every business more or less depends on marketing to get more customers, and reaching out to the right customer has always been a big challenge. Thanks to ML and AI, that problem is now better handled as it can anticipate the customer behavior to a great extent. The ML- and AI-based applications are being used to differentiate between potential prospects who are more likely to buy or subscribe to the offer or product and casual candidates. They are also being used to provide an absolute personalized offer to convert or retain the customers. A churn predictor is again used heavily to identify the group of consumers who are likely to discontinue the usage of any particular product or service. Advanced customer segmentation for hypertargeting is being done using ML and AI. Finally, a lot of marketing content is being generated artificially using ML and AI to send out the best-performing content.

Manufacturing and Automobile

The manufacturing domain has not been able to escape the wave of ML and AI either. The most predominant use is predictive maintenance as ML- and AI-based applications can help in preventing potential damages by predicting the need for maintenance in advance based on earlier data. Automobile companies are using telematics data to learn the driving patterns of the customers and act more promptly to help them in many ways. They are also using web data to understand their customers better to try to personalize the experience for seamless navigation during the online journey.

Social Media

Most people (the young generation in particular) spend a great deal of time on social media without realizing that a lot of the applications are using ML and AI. Facebook, YouTube, LinkedIn, Twitter, and other similar apps use ML heavily to provide the experience. From photo auto-tag suggestions to recommendations of friends, everything is driven by ML and AI. They are also used to generate subtitle and language translations for various platforms such as YouTube. Various search engines and voice assistants are using a good amount of ML implementation in them.

Others

There are many other applications where ML and AI are used. For example, email spam filters use ML instead of rule-based systems. One advantage that the ML approach offers over the traditional rule-based system is that the former automatically updates and upgrades itself as per the new mails to make this distinction. Another area is the oil and gas industry where ML and AI help in analyzing underground minerals and finding alternative energy sources. ML and AI are also being used in transportation as they can predict the likely traffic conditions and alert you in advance.

Challenges

So far, we have covered the capabilities and impact that ML and AI can have on this world. However, there are still lots of gaps that exist in order to realize the true potential of ML and AI. To start with, the shortage of skilled talent is a major blocker to the advancement of ML and AI. We have already discussed that people will require the combination of multiple skills to excel in this field, which makes it more difficult to find those kinds of resources.

"Finding a data scientist is hard. Finding people who understand who a data scientist is, is equally hard."

—Krzysztof Zawadzki

The next challenge is access to increased computing power. Although we have an availability of highly capable processing units as GPUs and TPUs, it's restricted to set of people rather than to everyone because of the cost factor and time taken to train big models. Hence, it still remains a challenge if the situation demands large data processing and model training. Security is the most critical aspect when it comes to using ML and AI as they use a lot of data to get trained in order to give better predictions. However, using personal and sensitive data for building the models can compromise users' data security and confidentiality. At the end of the day, machine learning and AI are not a silver bullet that can solve all problems. Another challenge associated with ML and AI is the explainability part as it is difficult to explain the rationale behind the predictions of the models. In fact, we can call them a black box as the interpretation of machine learning mapping features can get utterly complicated sometimes and might not make lot of sense to other stakeholders. There are certain areas where ML and AI cannot be applied, and hence many initiatives and applications are bound to fail.

Requirements

The following chapters makes use of Docker to build and deploy containers, and hence your system should have Docker installed and working properly. You also need to have admin rights to install some of the dependencies. You should also have a virtual box installed on your system. To deploy apps using a cloud service, you should have a Google Cloud account.

Conclusion

In this chapter, we went over the fundamentals of both machine learning and deep learning. We also saw the process of building a model on Databricks. We covered the different applications of machine learning and deep learning along with their existing challenges.

CHAPTER 2

Model Deployment and Challenges

You got a refresher on machine learning concepts in the previous chapter, so it is now logical to move to the next stage. What is machine learning deployment, and what are some of the common challenges when doing it?

This chapter covers two main themes. First, the chapter talks about what exactly model deployment is and the different aspects of productionalizing the model. Second, the chapter covers the different challenges that are faced during ML productionization. The challenges can be observed at both stages of model deployment (the pre-deployment and post-deployment phases), but for simplicity we are going focus on the set of challenges as a whole. Although the challenges faced by the machine learning team can be unique for each specific case, we will go over the most common cases pertaining to deployment.

© Pramod Singh 2021
P. Singh, *Deploy Machine Learning Models to Production*,
https://doi.org/10.1007/978-1-4842-6546-8_2

Model Deployment

In the previous chapter, we saw what it takes to build a machine learning model or a deep learning model (be it local or in the cloud). The level of complexity or the nature of the model can vary on a case-to-case basis, but the underlying framework remains similar. We have some typical data coming in from the data source (which can be a single source or multiple sources), followed by series of data cleaning and preprocessing steps, and then we extract or create important features from the input data to train the machine learning model for a specific use case. Once the model is trained or ready to be used in a production environment, it can be exposed to unseen data for making predictions through some APIs. However, the last part creates a handful of challenges. If we try to observe the cycle after building a successful ML model, it looks something like Figure 2-1.

The first stage is to deploy the trained ML model in production and test the results. This is followed by the performance monitoring of the model on continuous levels. Once the model starts performing below the expected benchmark level, the model needs to be retrained and evaluated again to replace the old model with the new. This includes the model management (versioning, features, etc.). The model is deployed again in production without affecting the existing user requests (you'll learn more about this in upcoming chapters on ML deployment). Once the new model is deployed, the same steps shown in Figure 2-1 are repeated through the framework.

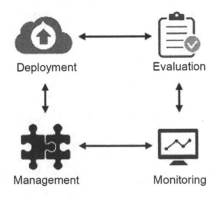

Figure 2-1. *ML model in deployment*

Although deployment is quite intuitive in nature because there are millions of applications out there that are already working seamlessly in the context of machine learning, it might need a bit of further explanation. Machine learning applications differ in nature compared to other typical software applications such as mobile applications in two main ways.

- Underlying model

- Underlying data

When we say we are going to deploy the model in production or productionalize the model, we are referring to integrating the machine learning model into an existing business application. In simple terms, we expose the ML model as REST API endpoints to serve the requests within the application platform or to direct user requests. This model to be deployed in production can be a stand-alone predictor that gives some output based on the algorithm used for batch data, or it can be used to serve requests in real time, making it a dynamic model. Model deployment can often be considered as the last stage in the machine learning cycle from a data scientist's standpoint; however, it's the start of what is known as the model management phase. Being able to successfully deploy a machine learning model requires lots of inputs and alignment from multiple stakeholders such as data scientists, data engineers, application developers, MLOps/DevOps, and business team members. To be honest, it's the most difficult stage in the machine learning lifecycle as a number of issues can crop up in the deployment phase. We will go into these challenges later in this chapter.

Why Do We Need Machine Learning Deployment?

By now we have a good understanding of what a machine learning model does and how it is trained, but the critical piece is to understand the role of the ML model in the overall business application. It can be simply a prediction or a combination of multiple predictions or recommendations of some sort. At the end of the day, it needs to make the overall business application more effective. For example, a ML model in production can be used to predict the propensity for each online visitor based on the activities on the website toward buying or not buying a specific product or anticipating the web traffic based on other factors. Irrespective of the actual role of ML in the overall application, deployment becomes an integral part to enable the ML model to talk to the application.

One might ask, why is it so important to take the ML model into production? There are multiple answers to this question, but the most important is to extract real value from the machine learning model. It has to become part of the application and power the application with all its predictions and insights. The best analogy that I can think of for not putting machine learning models into production is like training for a sports event but not participating in it. This restricts the impact that it could have had if embedded in the application. Having said that, in some cases, it makes more sense to not put models into production. As mentioned earlier, it depends on the case and the actual context of the application in which machine learning is used. In some cases, a stand-alone model does prove to be simpler and easier to use. Stand-alone models are much faster to build, train, make predictions, and extract insights for the business and do decision-making. However, this might not be a relevant approach in the following situations:

- The data is huge.
- The data is streaming.

- There are lots of active users.

- There are faster responses.

The ML models specifically trained on big data to handle a similar kind of data for predictions need to be well managed in the overall application. Similarly, if the data at hand is of a streaming type, then the model should be integrated with the application to handle the continuously incoming data. Another aspect is when you have a large number of users; in that case, you want to ensure that the models are able to handle that many requests and multiple instances of the same models are running in parallel to serve the requests. One key thing to remember is that deploying an ML model in production does not guarantee consistent quality in predictions. In reality, the performance of the model is often expected to deteriorate quickly as the model is exposed to real data (this effect is known as *drift*, and you'll learn more about it in the "Challenges" section), and that's where model management plays an important role.

To summarize, the deployment helps to extract the real value of machine learning by integrating the model with the application to generate tangible business insights. It also means that predictions can be made on real-time data.

Challenges

I wouldn't be writing this section of the chapter if it were easy to deploy ML models without any hiccups. It's rare that someone has the ideal conditions to go from the dev stage to production without some alterations and tweaks. We have to understand that there is a major difference between developing a model and deploying a model. The expectations in terms of performance, speed, and resource consumption are all different for both of these tasks. More often than not, there are two separate groups working on each of these stages separately (although this trend is changing as more data scientists are taking up the ML deployment task and as DevOps folks are learning to build ML/DL models on their own). Out of all the expectations from the ML models, the most differentiating one is that the performance of the model needs to be continuously monitored while the model is in production serving live requests from users because the application must have the best available version of the model in production.

The famous Google paper published by Sculley et al. in 2015, "Hidden Technical Debt in Machine Learning Systems," presented a different viewpoint to the machine learning community when it questioned the actual role and importance of machine learning in the overall application (https://papers.nips.cc/paper/5656-hidden-technical-debt-in-machine-learning-systems.pdf).

> *"They said that in real-world Machine Learning (ML) systems, only a small fraction is comprised of actual ML code. There is a vast array of surrounding infrastructure and processes to support their evolution."*

The paper mentioned that other components such as data dependencies, model complexity, reproducibility, testing, monitoring, and version changes play comparatively a bigger role in getting a realistic ML application. The general perception of the role of machine learning before that paper was that it was the most critical part of the overall lifecycle, as shown in Figure 2-2.

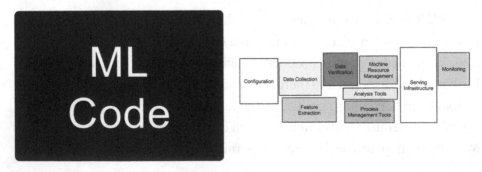

Figure 2-2. *ML role in application*

According to the paper, this is far from reality. In actuality, the machine learning code is a small piece of the entire scheme of things, as shown in Figure 2-3. There are multiple other components that drive the real value of an ML-based application.

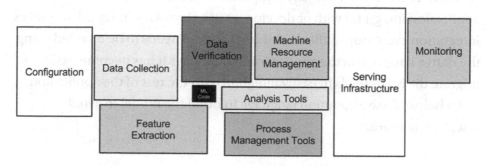

Figure 2-3. *Actual ML role in application*

To be honest, the challenges faced during the model deployment are closely related to the components shown in Figure 2-3 as they make the whole application practical. As mentioned earlier, there are other challenges that deployment teams can face. The following sections offer you a fair idea regarding what to expect when you take your ML model into production.

Challenge 1: Coordination Between Stakeholders

As mentioned earlier, the most natural hurdle while deploying a model is to align with other team members who do not have data science or machine learning backgrounds such as DevOps, application developers, and business team members. Model deployment is a group task that needs constant communication and a common understanding of the overall objective. Proper planning at the early stages of ML model development will help the MLOps/DevOps team to prepare well in advance for the deployment.

Challenge 2: Programming Language Discrepancy

It is likely that the machine learning model is built in a different language (Python/TensorFlow/PyTorch) compared to the language that the application integrates with or developers use (Java/C++/Ruby). This makes integration even more difficult as the ML model needs to be recoded using the native language of the application. These days it has become easier to migrate the ML models to easily integrate with the rest of the application, but it helps to use a common language to build the model to avoid integration issues.

Challenge 3: Model Drift

Model drift is a common phenomenon when it comes to the performance of machine learning models in production. Model drift happens when the prediction performance of the model degrades below the accepted benchmark. Again, it depends on what context the model is being used in and how those predictions are being evaluated, but every application would need to have the best version of the back-end ML model for higher efficiency and output. This becomes one of the main reasons to track the performance of the deployed model on an ongoing basis in production.

But if we look deeper at the reasons for performance degradation of the models in production, we can attribute degradation to a couple of reasons.

- Changing behavior of new data

- Changing interpretation of the new data

Changing Behavior of the Data

We would typically train the model on historical data with the necessary features and after a satisfied performance level (after all the hyperparameter tuning and testing) put the model into the production. However, the simple fact is that data can change in different ways, with time bringing new variations and dimensionality that were previously unseen by the model (during training). This impacts the performance of the model in real time or in production. An example is if machine learning was used to predict the likelihood or propensity for a user to convert or not (buy the product or not) based on the training using the previous two years of historical data that captured the online behavior (how the user navigates on a website before making a buying decision). At the start, the model is performing really well and able to decide among all the online visitors who were potentially serious buyers and who were casual visitors, but over a period of time, the buying pattern and behavior of the users change, and there are different outcomes than what the machine learning model suggested. Perhaps the model learned different decision boundaries to separate buyers from nonbuyers and now needs to be realigned per the new underlying patterns. This is known as *model drift* due to the dynamic nature of the incoming data on which the predictions are made.

Changing Interpretation of the New Data

In another case, the interpretation of the class or label might change over a period of time. For example, if earlier a given set of customers belonged to class A but due to a change in understanding and the business case, the category is either altered or combined with other category (let's say class C), that would impact the model's overall performance. We need to retrain the model with the new set of labels (for historical data) before putting it in production again. This is also known as *concept drift* in machine learning.

Challenge 4: On-Prem vs. Cloud-Based Deployment

Many businesses find it challenging to decide whether they need to go with model deployment using on-premise resources or to opt for cloud services. Both of these decisions can have a different impact in the long run based on the number of application users. The DS team and the infrastructure team should consider all the pros and cons before deciding on a productionization strategy.

Challenge 5: Clear Ownership

Ownership can sometimes become confusing as there needs to be clearly defined roles and responsibilities of each team member when it comes to model deployment. Data scientists assume their job is done once the model is built, whereas DevOps/developers have little clue of what's going on in the model and hence need more inputs to properly integrate the model in the application. At the end of the day, it's a joint responsibility of the entire team working on the deployment to ensure that the right model gets deployed into production and to avoid unnecessary delays.

Challenge 6: Model Performance Monitoring

Although monitoring comes after the model deployment stage, it still needs to be planned before the actual deployment. The framework needs to be in place to track the performance of the model on a consistent basis. This helps to formulate the strategy around the following questions:

- When to retrain the model again?
- How much data to use to retrain the model?
- What is the performance benchmark for the model?

Over a period of time, the data from this framework helps to come up with benchmark figures for the acceptable performance of the model in production.

Challenge 7: Release/Version Management

There are lots of back-and-forth activities when it comes to machine learning deployment, and version tracking becomes an important part of the overall successful process of deployment. Hence, version control helps to track which model is the best model, the different file dependencies, and the data resource pointers. There are multiple ways in which version control can be handled; Git is the most widely used one to track different versions and staged deployment.

Challenge 8: Privacy Preserving and Secure Model

Another challenge in model deployment is to safeguard the models from any adversarial attacks and expose the model data to third-party infiltrations. This becomes paramount to assure the model is not being exposed to any unauthorized user operating in a secure mode.

There are multiple other challenges as well such as scalability, exposing services, etc., and we are going to keep coming back to these challenges in the upcoming chapters. In the next chapter, we are going to look at different modes of model deployment. We can break model deployment into four different approaches.

- Deploying a model locally

- Saving the model on a production server

- Deploying the model as a REST service

- Managed services

Conclusion

In this chapter, we went over the fundamentals of model deployment and its associated challenges.

CHAPTER 3

Machine Learning Deployment as a Web Service

In this chapter, we are going to go over how to use different web frameworks for deploying machine learning and deep learning models as web services hosted on the local system. This chapter covers three main topics.

The chapter first introduces the Flask framework and how to deploy an ML model using it. Then the chapter shows how to build a standard machine learning model and deploy it using another web framework, Streamlit. Finally, the chapter covers how to deploy a trained deep learning model using the Streamlit platform again. If you are already comfortable with Flask basics, feel free to skip to the deployment part of the first section to understand the process of deploying machine learning models using Flask. As mentioned in the previous chapter, we mainly use Jupyter notebooks to develop and test the models locally, but when we want to connect our ML model to an app or a web service, we essentially need to deploy it using a web server. This web server can be hosted locally or in the cloud. There are many different ways in which a machine learning model can be deployed, but in this chapter, we are going to explore two methods: Flask and Streamlit.

© Pramod Singh 2021
P. Singh, *Deploy Machine Learning Models to Production*,
https://doi.org/10.1007/978-1-4842-6546-8_3

Introduction to Flask

In simple words, Flask is an open source lightweight web framework built in Python to deploy web applications. When we say *web framework*, we mean a group of resources needed to run a web application. This might include different modules, libraries, and tools that can be used by the web developer to successfully build and run the application. Unfortunately, this book does not do a deep dive into Flask, but for those of you who have never used it before, the following code snippet gives a quick introduction to Flask. To use Flask, we first need to install it on the local machine. We can simply use `pip install flask` to install Flask.

```
[In]: from flask import Flask
```

```
[In]: app = Flask(__name__)
```

```
[In]: @app.route("/")
[In]: def hello():
          return "Hello World!"
```

```
[In]: if __name__ == '__main__':
          app.run(debug=True)
```

route Function

The `route` function is a decorator that tells which URL is associated with a particular function. It has two parameters.

- `rule`
- `options`

The `rule` indicates the URL path and its binding with the given function. It renders the current output of the function when the URL is opened in the browser. The options let you pass different sets of parameters.

run Method

The run method executes the application to run on the particular web server. It has four parameters that can be passed during execution. All of these are optional parameters, and app.run can be executed without passing any of them as well.

- host
- port
- debug
- options

By default, the application runs on localhost (127.0.0.1). The debug option is set to false by default and can be set to true to see the debug information.

Deploying a Machine Learning Model as a REST Service

Now that we know how the Flask framework works, we are going to build a simple linear regression model and deploy it using the Flask server. We start by importing the required libraries.

```
[In]: import pandas as pd
[In]: import numpy as np
[In]: from sklearn.linear_model import LinearRegression
[In]: import joblib
```

We load the dataset into pandas, and as we can see, our dataset contains five input columns and a target column.

```
[In]: df=pd.read_csv('Linear_regression_dataset.
      csv',header='infer')
[In]: df.sample(5)
[Out]:
```

	var_1	var_2	var_3	var_4	var_5	output
876	664	724	79	0.331	0.264	0.361
1167	654	667	85	0.313	0.250	0.384
724	729	688	84	0.318	0.253	0.424
1222	842	697	102	0.337	0.268	0.400
482	772	771	76	0.320	0.251	0.401

Since the idea is not to build a super-powerful model but rather deploy an ML model, we need not split this data into train and test sets. We fit a linear regression model and get a decent r-square value.

```
[In]: X=df.loc[:,df.columns !='output']
[In]: y=df['output']
[In]: lr = LinearRegression().fit(X, y)
[In]: lr.score(X,y)
[Out]: 0.8692670151914198
```

The next step is to save the trained model that can be loaded back while serving it as a web service. We make use of the joblib library that serializes the model (saves coefficient values for input variables as a dictionary).

```
[In]: joblib.dump(lr,'inear_regression_model.pkl')
```

Now that we have saved the model, we can create the main app.py file, which will spin up the Flask server to run the ML model as a web app.

```
[In]: import pandas as pd
[In]: import numpy as np
[In]: import sklearn
[In]: import joblib
[In]: from flask import Flask,render_template,request
[In]: app=Flask(__name__)

[In]: @app.route('/')
```

```
[In]: def home():
            return render_template('home.html')

[In]: @app.route('/predict',methods=['GET','POST'])

[In]: def predict():
      if request.method =='POST':
            print(request.form.get('var_1'))
            print(request.form.get('var_2'))
            print(request.form.get('var_3'))
            print(request.form.get('var_4'))
            print(request.form.get('var_5'))
            try:
                  var_1=float(request.form['var_1'])
                  var_2=float(request.form['var_2'])
                  var_3=float(request.form['var_3'])
                  var_4=float(request.form['var_4'])
                  var_5=float(request.form['var_5'])
                  pred_args=[var_1,var_2,var_3,var_4,var_5]
                  pred_arr=np.array(pred_args)
                  preds=pred_arr.reshape(1,-1)
                  model=open("linear_regression_model.
                  pkl","rb")
                  lr_model=joblib.load(model)
                  model_prediction=lr_model.predict(preds)
                  model_prediction=round(float(model_
                  prediction),2)
            except ValueError:
                  return "Please Enter valid values"
      return render_template('predict.html',prediction=model_
      prediction)
[In]: if __name__=='__main__':
            app.run(host='0.0.0.0')
```

Let's go over the steps to understand the details of the app.py file. First, we import all the required libraries from Python. Next, we create our first function, which is the home page that renders the HTML template to allow the users to fill input values. The next function is publishing the predictions by the model on those input values provided by the user. We save the input values into five different variables coming from the user and create a list (pred_args). We then convert that into a numpy array. We reshape it into the desired form to be able to make a prediction on it. The next step is to load the trained model (linear_regression_model.pkl) and make the predictions. We save the final output into a variable (model_prediction). We then publish these results via another HTML template, predict.html. If we run the main file (app.py) now in the terminal, we would see the page come up asking the user to fill the values, as shown in Figure 3-1.

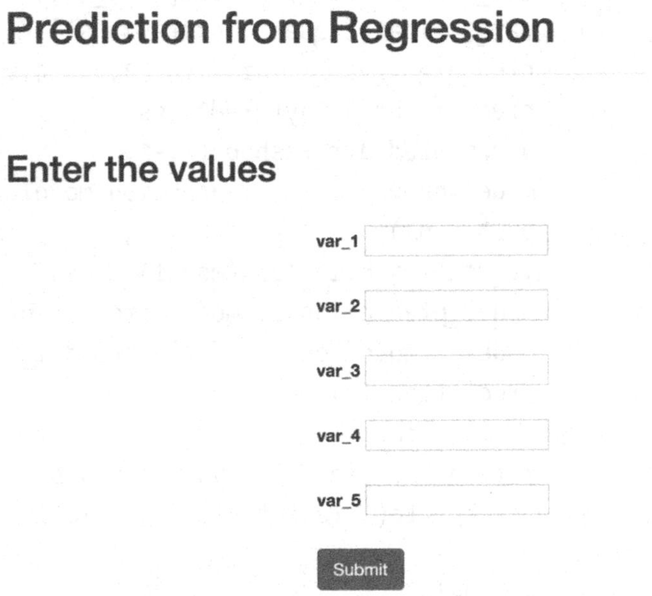

Prediction from Regression

Enter the values

var_1

var_2

var_3

var_4

var_5

Submit

Figure 3-1. *Input for ML prediction*

Templates

There are two web pages that we have to design to post requests to the server and receive the response message that is the prediction by the ML model for that particular request. Since this book doesn't focus on HTML, you can simply use these files as is without making any changes to them, as shown in Figure 3-2. But for curious readers, we are creating a form to request five values in five different variables. We are using a standard CSS template with some basic fields. Users with prior knowledge of HTML can feel free to redesign the home page as per their requirements.

```html
<!DOCTYPE html>
<html>
<head>
  <<link rel="stylesheet" href="https://maxcdn.bootstrapcdn.com/
  bootstrap/3.3.7/css/bootstrap.min.css" integrity="
  sha384-BVYiiSIFeK1dGmJRAkycuHAHRg320mUcww7on3RYdg4Va+PmSTsz/K68vbdEjh4u"
  crossorigin="anonymous">

  <title>Prediction Using Flask</title>
  </head>

  <body>
      <h1><div style="text-align:centre"><font color='blue'>Prediction from
      Regression </font></div></h1>
      <hr>
      </br>

      <h2><div style="text-align:centre">Enter the values </div></h2>

      </br>

      <div class='container'>
          <div class='row'>
              <div class='col-6'>
                  <form method="POST" action="/predict">
                      <div class='form-group'>
                          <label for="var_1"><p class="font-weight-bold">
                          var_1</p></label>
                          <input type='text' name="var_1">
                          </div>

                      <div class='form-group'>
                          <label for="var_2"><p class="font-weight-bold">
                          var_2</p></label>
                          <input type='text' name="var_2">

                      </div>
                      <div class='form-group'>
                          <label for="var_3"><p class="
                          font-weight-bold">var_3</p></label>
                          <input type='text' name="var_3">

                      </div>
                      <div class='form-group'>
                          <label for="var_4"><p class="
                          font-weight-bold">var_4</p></label>
                          <input type='text' name="var_4">

                      </div>
                      <div class='form-group'>
                          <label for="var_5"><p class="
                          font-weight-bold">var_5</p></label>
                          <input type='text' name="var_5">

                      </div>

                      <input class='btn btn-primary' type
                      ='submit' value='Submit'>

                  </form>
              </div>
              </div>
              </div>
          </div>

      </body>
</html>
```

Figure 3-2. *Input request HTML form*

The next template is to publish the model prediction to the user. This is less complicated compared to the first template as there is just one value that we have to post back to the user, as shown in Figure 3-3.

```html
<!DOCTYPE html>
<html>
<head>
    <<link rel="stylesheet" href="https://maxcdn.bootstrapcdn.com/
    bootstrap/3.3.7/css/bootstrap.min.css" integrity="
    sha384-BVYiiSIFeK1dGmJRAkycuHAHRg320mUcww7on3RYdg4Va+PmSTsz/K68vbdEjh4u"
    crossorigin="anonymous">

    <title>Prediction by ML model</title>
    </head>

    <body>
        <h1><div style="text-align:center"><font color='blue'>Prediction
        Result </font></div></h1>
        <hr>
        <div class='card text-center' style="width:21.5em;margin:0 auto;">
            <div class="card-body">
                <p class="card text"><h1><font color='blue'>{{prediction}}</
                font></h1></p>
                </div>
                </div>
        </body>
</html>
```

Figure 3-3. *Model prediction HTML form*

Let's go ahead and input values for the model prediction, as shown in Figure 3-4. As we can observe in Figure 3-5, the model prediction result is a continuous variable since we have trained a regression model.

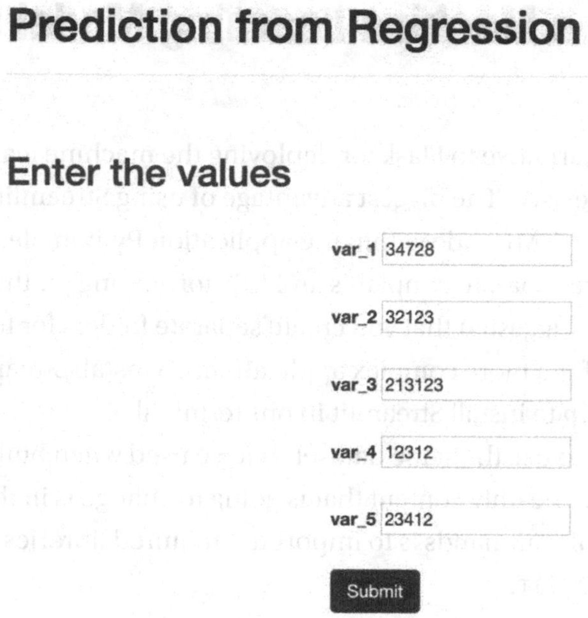

Prediction from Regression

Enter the values

var_1 34728

var_2 32123

var_3 213123

var_4 12312

var_5 23412

Submit

Figure 3-4. *User inputs page*

Prediction Result

3608.45

Figure 3-5. *Model prediction*

As we saw, Flask makes it easy to deploy the machine learning app as a web service. One disadvantage of using Flask is that since it is a lightweight web framework, it has a limited capacity to handle complex applications. Another disadvantage is that most data scientists are not comfortable with using HTML and JavaScript to create the front end for the application. Hence, in the next section, we are going to look at a much simpler alternative of deploying a machine learning app using Streamlit. This makes it easier to develop a simple UI for the app compared to Flask.

Deploying a Machine Learning Model Using Streamlit

Streamlit is an alternative to Flask for deploying the machine learning model as a web service. The biggest advantage of using Streamlit is that it allows you to use HTML code within the application Python file. It doesn't essentially require separate templates and CSS formatting for the front-end UI. However, it is suggested that you create separate folders for templates and style guides for a more complex application. To install Streamlit, we can simply use pip to install Streamlit in our terminal.

We are going to use the same dataset that we used when building the Flask app model. The only content that is going to change is in the app.py file. The first set of commands is to import the required libraries such as joblib and streamlit.

```
[In]: import pandas as pd
[In]: import numpy as np
[In]: import joblib
[In]: import streamlit
```

In the next step, we import the trained linear regression model to be able to predict on the test data.

```
[In]: model=open("linear_regression_model.pkl","rb")
[In]: lr_model=joblib.load(model)
```

The next step is to define a function to make predictions using the trained model. We pass the five input parameters in the function and do a bit of reshaping and data casting to ensure consistency for predictions. Then we create a variable to save the model predictions result and return it to the user.

```
[In]: def lr_prediction(var_1,var_2,var_3,var_4,var_5):
          pred_arr=np.array([var_1,var_2,var_3,var_4,var_5])
```

```
preds=pred_arr.reshape(1,-1)
preds=preds.astype(int)
model_prediction=lr_model.predict(preds)
    return model_prediction
```

In the next step, we create the most important function. We accept the user input from the browser and render the model's final predictions on the web page. We can name this function anything. For example, I have used `run` (since it does the same thing as Flask's `app.run`). In this function, we include the front-end code as well such as defining the title, theme, color, background, etc. For simplicity purposes, I have kept it basic, but this can have multiple levels of enhancements. For more details, you can visit the Streamlit website. Next, we create five input variables to accept the user input values from the browser. This is done using Streamlit's `text_input` capability. The final part contains the model prediction, which gets the input from our `lr_prediction` function defined earlier and gets rendered in the browser through `streamlit.button`.

```
[In]: def run():
            streamlit.title("Linear Regression Model")
            html_temp="""

            """
            streamlit.markdown(html_temp)
 var_1=streamlit.text_input("Variable 1")
 var_2=streamlit.text_input("Variable 2")
 var_3=streamlit.text_input("Variable 3")
 var_4=streamlit.text_input("Variable 4")
 var_5=streamlit.text_input("Variable 5")
```

```
            prediction=""

        if streamlit.button("Predict"):
                prediction=lr_prediction(var_1,var_2,var_3,
                var_4,var_5)
        streamlit.success("The prediction by Model : {}".
        format(prediction))
```

Now that we have all the steps mentioned in the application file, we can call the main function (run in our case) and use the streamlit run command to run the app.

```
[In]: if __name__=='__main__':
        run()
[In]: streamlit run app.py
```

Once we run the previous command, we will soon see the app up and running on port 8501, as shown in Figure 3-6. We can simply click the link and access the app.

You can now view your Streamlit app in your browser.

Local URL: **http://localhost:8501**
Network URL: **http://192.168.1.4:8501**

Figure 3-6. *Access running app*

Once we are at http://localhost:8501/, we will see the screen shown in Figure 3-7.

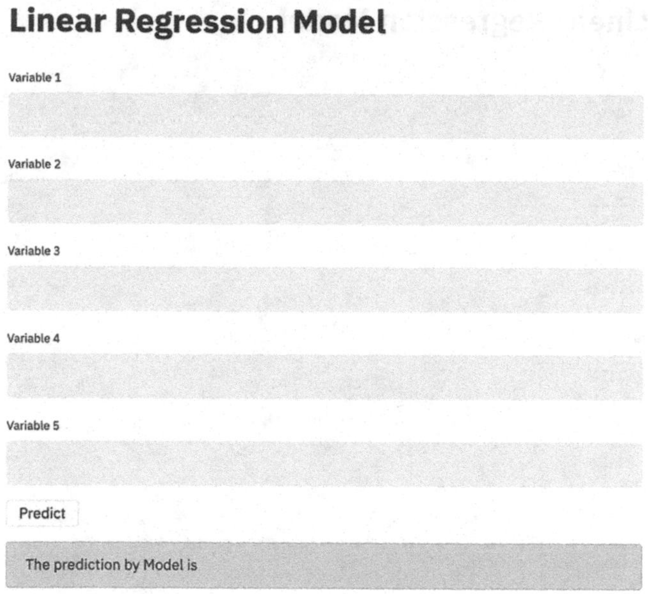

Linear Regression Model

Variable 1

Variable 2

Variable 3

Variable 4

Variable 5

Predict

The prediction by Model is

Figure 3-7. User input page

As you can see, we have nothing fancy here in the UI, but it serves the overall purpose for us to be able to interact with the model/app behind the scenes. It is similar to what we had built using Flask, but it needed much less HTML and CSS code. We can now go ahead and fill in the values, as shown in Figure 3-8, and get the model prediction. For comparison sake, we fill in the same values that we filled in the Flask-based app.

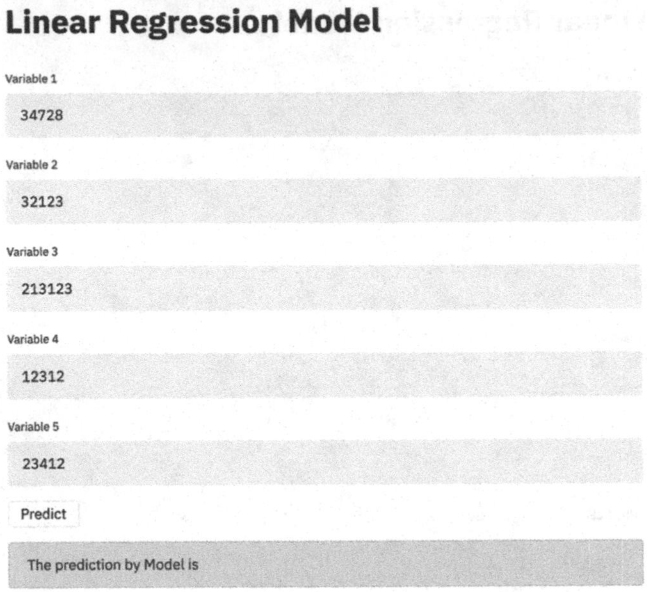

Linear Regression Model

Variable 1

34728

Variable 2

32123

Variable 3

213123

Variable 4

12312

Variable 5

23412

Predict

The prediction by Model is

Figure 3-8. *Providing input values for the model*

After filling in the values, we need to click the Predict button to fetch the model prediction result and *voilà*—it's the same number that we got in the Flask-based app, as shown in Figure 3-9.

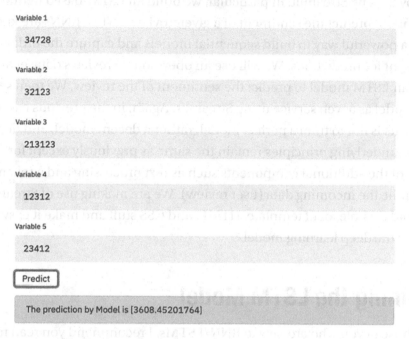

Figure 3-9. *Model prediction*

Now that we have seen how to deploy a traditional machine learning model using Flask and Streamlit, we can move on to the last topic of this chapter, which focuses on deploying a deep learning model (LSTM) as a web service.

Deploying a Deep Learning Model

In the previous sections of the chapter, we covered the process to build and deploy a machine learning model (linear regression) using two different frameworks.

- Flask
- Streamlit

In this section, we will see how to build a deep learning model and deploy it using Streamlit. In particular, we build an LSTM-based neural network to predict the sentiment of a given review. LSTM/RNNs are known to be a powerful way to build sequential models and capture the order of the input for predictions. We will use an open source reviews dataset and train an LSTM model to predict the sentiment of the review. We then serve this model as a web service using Streamlit. Again, in the previous cases, the focus is not to train a perfect model, rather a decent model, and deploy it. The underlying principles remain the same as previously except for some of the additional components such as text processing and a tokenizer to handle the incoming data (user review). We are making use of Streamlit to avoid a whole lot of template, HTML, and CSS stuff and make it easy to deploy the deep learning model.

Training the LSTM Model

For those of you who are new to RNN/LSTMs, I recommend you read more about them to understand how they work and when they can be used for predictions. The first step is to import all the required libraries.

```
[In]: import numpy as np
[In]: import pandas as pd
[In]: from tensorflow.keras.models import Sequential
[In]: from tensorflow.keras.layers import LSTM,Embedding
[In]: from tensorflow.keras.layers import Dense
[In]: from tensorflow.keras.preprocessing.text import Tokenizer
[In]: from tensorflow.keras.preprocessing.sequence import pad_
       sequences
[In]: from sklearn.model_selection import train_test_split
[In]: from keras.utils.np_utils import to_categorical
[In]: import re
[In]: import pickle
```

The next step is to load the data and check the size of the data.

```
[In]: df = pd.read_csv('reviews_dataset.tsv.zip',header=0,
delimiter="\t", quoting=3)
```

```
[In]: df = df[['review','sentiment']]
```

As we can see, we have 25,000 records in our dataset and two equal categories of sentiments.

```
[In]: df.shape
[Out]: (25000, 2)
```

```
[In]: df.sentiment.value_counts()
[Out]:
1    12500
0    12500
```

Next, we apply bit of text preprocessing to clean the reviews using regular expressions.

```
[In]: df['review'] = df['review'].apply(lambda x: x.lower())
[In]: df['review'] = df['review'].apply((lambda x: re.sub('[^a-zA-z0-9\s]','',x)))
```

We restrict the number of features to 1,000 and tokenize the reviews and add padding to make each review the same size.

```
[In]: max_features = 1000
[In]: tokenizer = Tokenizer(num_words=max_features, split=' ')
[In]: tokenizer.fit_on_texts(df['review'].values)
[In]: X = tokenizer.texts_to_sequences(df['review'].values)
[In]: X = pad_sequences(X)
[In]: X.shape

(25000, 1473)
```

We then keep the embedding layer size as 50.

```
[In]: embed_dim = 50

[In]: model = Sequential()
[In]: model.add(Embedding(max_features, embed_dim,
      input_length = X.shape[1]))
[In]: model.add(LSTM(10))
[In]: model.add(Dense(2,activation='softmax'))
[In]: model.compile(loss = 'categorical_crossentropy',
      optimizer='adam',metrics = ['accuracy'])
[In]: print(model.summary())
[Out]:
```

Model: "sequential"

Layer (type)	Output Shape	Param #
embedding (Embedding)	(None, 1473, 50)	50000
lstm (LSTM)	(None, 10)	2440
dense (Dense)	(None, 2)	22

```
Total params: 52,462
Trainable params: 52,462
Non-trainable params: 0
```

```
[In]: y = pd.get_dummies(df['sentiment']).values
[In]: X_train, X_test, y_train, y_test = train_test_split
      (X,y, test_size = 0.25, random_state = 99)
[In]: print(X_train.shape,y_train.shape)
[In]: print(X_test.shape,y_test.shape)

[Out]:
(18750, 1473) (18750, 2)
(6250, 1473) (6250, 2)
```

```
[In]: model.fit(X_train, y_train, epochs = 5, verbose = 1)
```

```
Train on 18750 samples
Epoch 1/5
18750/18750 [==============================] - 263s 14ms/sample - loss: 0.4837 - accuracy: 0.7610
Epoch 2/5
18750/18750 [==============================] - 278s 15ms/sample - loss: 0.3485 - accuracy: 0.8545
Epoch 3/5
18750/18750 [==============================] - 273s 15ms/sample - loss: 0.3279 - accuracy: 0.8623
Epoch 4/5
18750/18750 [==============================] - 274s 15ms/sample - loss: 0.3077 - accuracy: 0.8709
Epoch 5/5
18750/18750 [==============================] - 270s 14ms/sample - loss: 0.2922 - accuracy: 0.8765
```

Now that we have trained the LSTM model, let's try to pass a test review to see the predictions by the model.

```
[In]: test = ['Movie was pathetic']
[In]: test = tokenizer.texts_to_sequences(test)
[In]: test = pad_sequences(test, maxlen=X.shape[1],
        dtype='int32', value=0)
[In]: print(test.shape)
[In]: sentiment = model.predict(test)[0]
        if(np.argmax(sentiment) == 0):
            print("Negative")
        elif (np.argmax(sentiment) == 1):
            print("Positive")
[Out]: Negative
```

As we can see, the model is able to predict well on the test review. The next step is to save the model and tokenizer using pickle and load it later for making predictions on user input reviews.

```
[In]: with open('tokenizer.pickle', 'wb') as tk:
        pickle.dump(tokenizer, tk, protocol=pickle.HIGHEST_
        PROTOCOL)
```

```
[In]: model_json = model.to_json()
        with open("model.json", "w") as js:
js.write(model_json)
```

```
[In]: model.save_weights("model.h5")
```

Now that we have saved the trained model and tokenizer, we can create the application script similar to the earlier app.py script.

```
[In]: import os
[In]: import numpy as np
[In]: import pandas as pd
[In]: import pickle
[In]: import tensorflow
[In]: from tensorflow.keras.preprocessing.text import Tokenizer
[In]: from tensorflow.keras.preprocessing.sequence import pad_
       sequences
[In]: import tensorflow.keras.models
[In]: from tensorflow.keras.models import model_from_json
[In]: import streamlit
[In]: import re
[In]: os.environ['TF_CPP_MIN_LOG_LEVEL'] = '2'
```

The step after importing the required libraries is to load the tokenizer and deep learning model.

```
[In]: with open('tokenizer.pickle', 'rb') as tk:
          tokenizer = pickle.load(tk)

[In]: json_file = open('model.json','r')
[In]: loaded_model_json = json_file.read()
[In]: json_file.close()
[In]: lstm_model = model_from_json(loaded_model_json)

[In]: lstm_model.load_weights("model.h5")
```

Next, we create a helper function to clean the input review, tokenize it, and pad the sequence. Once it's converted into numerical form, we will use the loaded LSTM model to make the sentiment prediction.

```
[In]: def sentiment_prediction(review):
    sentiment=[]
input_review = [review]
input_review = [x.lower() for x in input_review]
input_review = [re.sub('[^a-zA-z0-9\s]','',x) for x in input_
review]

input_feature = tokenizer.texts_to_sequences(input_review)
input_feature = pad_sequences(input_feature,1473,
padding='pre')
    sentiment = lstm_model.predict(input_feature)[0]

    if(np.argmax(sentiment) == 0):
pred="Negative"
    else:
pred= "Positive"

    return pred
```

At the end, we create the run function to load the HTML page and accept the user input using Streamlit functionality (similar to the earlier model deployment). The only difference is that instead of loading multiple inputs, this time we load just a single review. We then pass this review to the sentiment prediction function created earlier.

```
[In]: def run():
streamlit.title("Sentiment Analysis - LSTM Model")
html_temp="""

    """

streamlit.markdown(html_temp)
    review=streamlit.text_input("Enter the Review ")
    prediction=""
```

```
     if streamlit.button("Predict Sentiment"):
         prediction=sentiment_prediction(review)
streamlit.success("The sentiment predicted by Model : {}".
format(prediction))
```

```
[In]: if __name__=='__main__':
        run()
```

Once we run the app.py file using the streamlit run command in the terminal, we can see the web service is running on localhost port 8501, as shown in Figure 3-10.

```
[In]: streamlit run app.py
```

You can now view your Streamlit app in your browser.

Local URL: **http://localhost:8501**
Network URL: **http://192.168.1.4:8501**

Figure 3-10. *Accessing the ML app*

We can access the ML service on port 8501 and can see the ML app running successfully. The page contains three things.

- Placeholder to provide a user review

- A Predict Sentiment button (to make a prediction using the model)

- The final result by the model

Let's provide a review in the input box and check the model predictions, as shown in Figure 3-11. First, we provide a positive review and click the Predict Sentiment button, as shown in Figure 3-12.

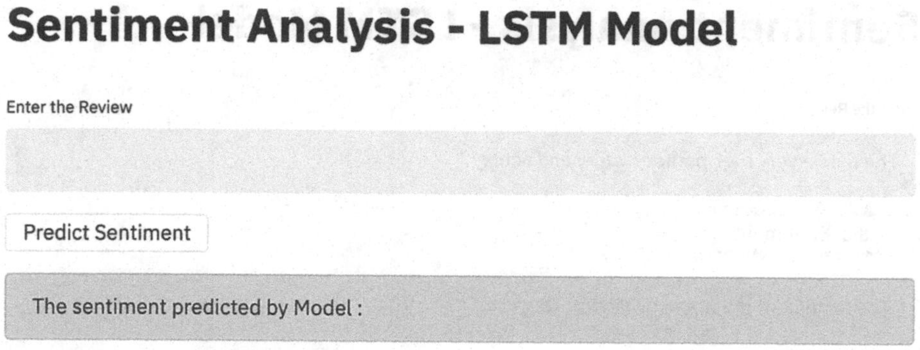

Sentiment Analysis - LSTM Model

Enter the Review

Predict Sentiment

The sentiment predicted by Model :

Figure 3-11. *User input page*

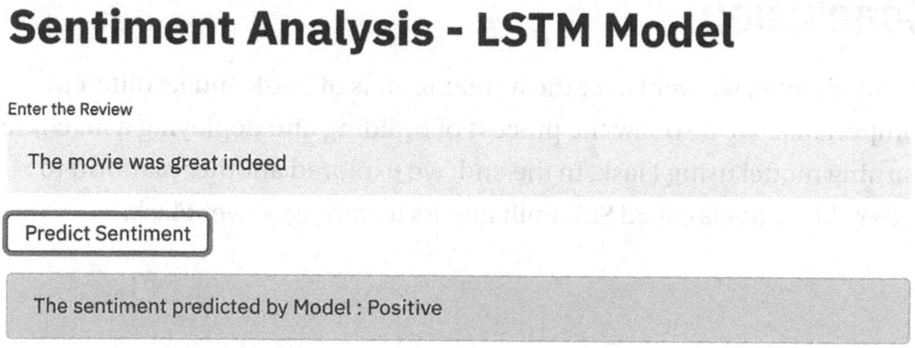

Sentiment Analysis - LSTM Model

Enter the Review

The movie was great indeed

Predict Sentiment

The sentiment predicted by Model : Positive

Figure 3-12. *Positive review prediction*

As we can observe, the outcome of the model prediction seems correct as it also predicts it to be a positive review. We try with another review—negative this time—and test for the model prediction, as shown in Figure 3-13. For the second review as well, we get a correct prediction by the model. We can replace the existing model with a much more powerful, well-trained, and optimized model to have better predictions.

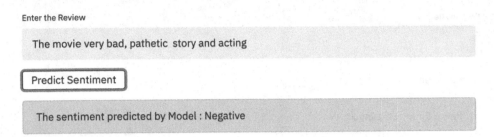

Sentiment Analysis - LSTM Model

Enter the Review

The movie very bad, pathetic story and acting

Predict Sentiment

The sentiment predicted by Model : Negative

Figure 3-13. Negative review prediction

Conclusion

In this chapter, we went over the fundamentals of Flask and its different components. We also saw the process of building and deploying a machine learning model using Flask. In the end, we explored another platform to deploy ML models called Streamlit and its advantages over Flask.

CHAPTER 4

Machine Learning Deployment Using Docker

Over the last few years, Docker has changed the way applications are deployed in production. Application architectures have moved from monolithic to microservices, with more control of continuous (ongoing) deployments that don't impact a large of part of the running applications. Docker has proved to be instrumental in allowing applications to run at scale and to be available all the time. Though it's been more than seven years since Docker was released, it's gotten a lot of attention from the developer community recently (especially by DevOps and MLOps teams). Companies large and small are using Docker in applications.

In this chapter, we will go over the Docker ecosystem and look at the different components of Docker. This chapter covers three main topics. First, the chapter covers what Docker is and why it is useful. Second, the chapter covers different components such as Docker images, the Docker server, Docker Hub, and containers. Finally, the chapter covers how to "Dockerize" an entire ML app and run it using a Docker container. Docker has grown hugely popular over the years, and there are many books dedicated to this subject only. We will cover the topics that are most relevant for us from a machine learning standpoint.

© Pramod Singh 2021
P. Singh, *Deploy Machine Learning Models to Production*,
https://doi.org/10.1007/978-1-4842-6546-8_4

What Is Docker, and Why Do We Need It?

Scenario 1: Imagine we want to install a new application on our laptop or host system. It is a straightforward process. We would typically download the installer file from the source of the application and run it in our system. Sometimes it installs perfectly on the first attempt; however, other times, we might run into different issues (dependencies, compatibility errors, etc.), as shown in Figure 4-1.

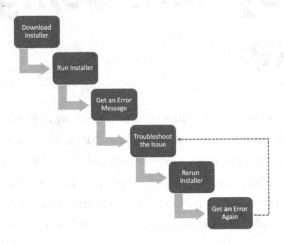

Figure 4-1. *Typical application installation*

Scenario 2: Imagine we have written some code on our machine using certain libraries within a particular environment. Everything runs perfectly when we execute it locally, but the moment we share it with someone else for testing, it breaks. This happens because of the change in the underlying configuration. It might be because of missing dependencies, different OSs, or some other reason.

Docker can handle both of the previous scenarios well as Docker ensures the same environment everywhere to run the application. If we are able to run an application or certain code on a particular machine, we can easily run it anywhere else as well. That's the beauty of Docker.

It makes it easy to install and run any application or program without having to worry about a set of dependencies and files on the underlying system. To elaborate with an example, if we have to install Redis (an in-memory database) on our system, we have to go to the Redis website and follow certain instructions.

```
$ wget http://download.redis.io/releases/redis-6.0.3.tar.gz
$ tar xzf redis-6.0.3.tar.gz
$ cd redis-6.0.3
$ make
```

From the output, it seems pretty straightforward, but at every stage we might run into an error because of missing dependencies to install Redis. Docker makes this super easy and fast. With a single line of code, we are able to install Redis on our system (within Docker).

```
docker container run redis
```

Introduction to Docker

"Docker is a computer program that performs operating-system-level virtualization, also known as containerization."

Containers allow us to package all the required files that our application needs such as libraries, binaries, and other dependencies within a single package. In this way, our application can be run on any machine and have the same behavior. The core idea is to be able to replicate the behavior of the application irrespective of the host on which it's running. For example, when we build any application, we typically write the code in the dev environment and later test it in the test environment. These environments generally vary from each other with respect to the production environment, and hence developers run into multiple issues due to different operating platforms. Docker ensures that if the application is successfully running in the dev environment, it can be safely deployed in the production environment as well, as shown in Figure 4-2.

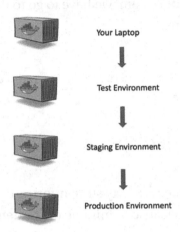

Your Laptop

Test Environment

Staging Environment

Production Environment

Figure 4-2. *Application deployment stages*

Docker vs. Virtual Machines

It's likely that you have used a virtual machine at some point. Virtual machines are powerful when we want to create a different environment and run separate applications using a different OS. There is a hypervisor that enables the virtualization layer to create VMs. Once virtual machines are created on the host system, each VM would have its own set of OSs, libraries, binaries, and apps. The VM would consume a certain amount of memory and hardware resources that it can't share with other VMs, as shown in Figure 4-3.

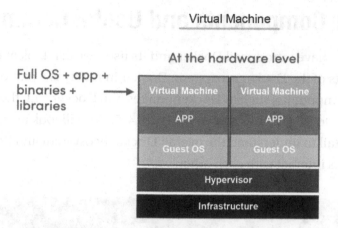

Figure 4-3. *Hypervisor-based virtualization*

In Docker there is no hypervisor; instead, there is a Docker server that isolates each app running in individual containers, as shown in Figure 4-4. Docker apps at the end of the day are lightweight virtual machines with the Linux OS; the advantage of this is that Docker only makes use of the memory and hardware resources as per the requirements, whereas in VMs we have to dedicate resources prior to the creation of the VM, and the resources can't be shared between VMs. This is a waste of resources and increases the cost (on the cloud).

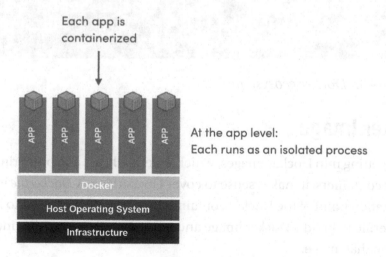

Figure 4-4. *Docker-based virtualization*

Docker Components and Useful Commands

Now that we have introduced Docker and its uses, we can look at the other components of the Docker ecosystem. Docker in reality is a platform with multiple components such as Docker images, the Docker CLI, the Docker server, and Docker Hub, as shown in Figure 4-5. We will look at each of these in detail to understand the overall Docker ecosystem and how the components interact with each other.

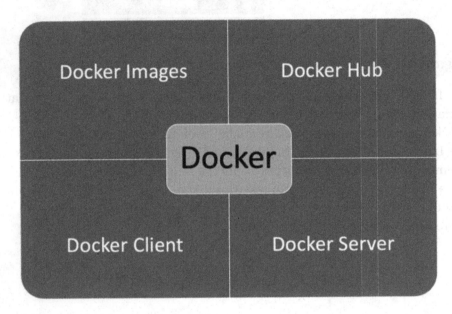

Figure 4-5. *Docker ecosystem*

Docker Image

Before getting into Docker images, which essentially act as a blueprint for Docker containers, it makes sense to cover Dockerfiles. A *Dockerfile* is the initialization point of the Docker container lifecycle. We first need to have a Dockerfile to build a Docker image and subsequently run a container based on that image.

Dockerfile

There are two types of scenarios that can arise when we work with Docker images.

- Using public Docker images

- Building customer Docker images

In the first case, we might not need to create a Dockerfile as we can simply pull the image from Docker Hub and create/run a container out of it without making any changes to the image. For example, if we want to run Nginx (a web server) using Docker, we simply pull the Nginx Docker image and run it without any changes. However, in the second case, where we want to customize the image as per our specific requirements, we need to create a Dockerfile.

A Dockerfile is a simple text file without any extension with the fixed name Dockerfile. It contains a bunch of commands (which can vary from file to file depending on the requirements) that get executed to build the Docker image. The process to create a Dockerfile is pretty straightforward, as shown in Figure 4-6. We will look at a Dockerfile to understand these commands.

Figure 4-6. *Dockerfile process*

We start with a base image that we want to leverage instead of starting from scratch. After that, we have additional commands related to files, dependencies, and environment variables. At the end, the Dockerfile contains the startup command that should be executed first upon running the container.

Dockerfile Commands

Although there are many commands (https://kapeli.com/cheat_
sheets/Dockerfile.docset/Contents/Resources/Documents/index)
that can be used inside the Dockerfile, most of the time the following
commands will suffice for the basic requirements. The idea is to start with
a base image for the new image and add dependencies and requirements
to the new image.

The following commands are used to create a Dockerfile. We will go
over these with the help of an example where we create a Dockerfile to run
a Node.js application.

- FROM

- COPY

- WORKDIR

- EXPOSE

- RUN

- CMD OR ENTRYPOINT

The following command provides the base image for the Dockerfile
that needs to be the starting point for rest of the image:

```
FROM node:alpine
```

In the case of the Node.js application, we mention the official Node.js
base image. FROM is always the first command in the Dockerfile because it
becomes the base layer for the entire container later.

The next command is to add or copy the contents of the working
directory or project folder from the host system to the Docker container
folder:

```
COPY . /app/usr/nodejs
```

We can use the COPY command to either create a new folder inside Docker or copy all the files and scripts inside the same working folder as well. In the previous case, we are copying all the application files to a new folder called nodejs inside the Docker container. The core idea is to provide all the application files inside Docker to be available while runtime.

Now that we have copied all the content in the Docker folder, it makes sense to run the application from that particular folder: nodejs. Hence, we change the working directory to the nodejs folder inside the Docker container. We make use of WORKDIR to set the directory of the app.

```
WORKDIR /app/usr/nodejs
```

The Docker container has its own set of network ports, and hence to respond to the incoming request, we need to expose the app on a port. The EXPOSE command allows us to make the app accessible on a given port to the external requests.

```
EXPOSE 5000
```

The RUN command helps us to install a set of dependencies and libraries to run the app inside the container. Instead of installing each dependency separately, we make use of a requirement.txt file containing all the required files with particular versions.

```
RUN pip install -r requirement.txt
```

The last command in Dockerfile is CMD, which is the startup command for the container.

```
CMD ["yarn", "start"]
```

Docker Hub

Docker Hub (https://hub.docker.com/) is for images what GitHub is for code repositories. It's a collection of public and private images. It contains official images as well as customized public images for different applications. For example, we get official images of Postgres, Spark, Ubuntu, and other such applications. It allows users to directly run a container using an official image and use the application. We need to have an account to access Docker Hub. It also allows us to pull and push Docker images from the local system. Once saved on Docker Hub, we can use a particular image at any point of time in any given environment. Now that the Dockerfile is created and we have seen Docker Hub as well, we can look into the details of the Docker client and server and use them to create a Docker image from the Dockerfile that we created earlier.

Docker Client and Docker Server

Docker has many components in its ecosystem; two of the most important are as follows:

- Docker client (CLI)

- Docker server (daemon)

Docker CLI is the gateway to interact with Docker on any system. We can issue instruction or commands to a Docker server through the CLI, and it communicates with them over the Docker server with slight changes. Docker CLI is also where we can see the output from the container or log in to the running container. To see more information about the Docker client and server, we can simply run docker version in our terminal, as shown in Figure 4-7.

```
(base) NDLMAC-49689:~ 49689$ docker version
Client: Docker Engine - Community
 Version:           19.03.8
 API version:       1.40
 Go version:        go1.12.17
 Git commit:        afacb8b
 Built:             Wed Mar 11 01:21:11 2020
 OS/Arch:           darwin/amd64
 Experimental:      false

Server: Docker Engine - Community
 Engine:
  Version:          19.03.8
  API version:      1.40 (minimum version 1.12)
  Go version:       go1.12.17
  Git commit:       afacb8b
  Built:            Wed Mar 11 01:29:16 2020
  OS/Arch:          linux/amd64
  Experimental:     false
 containerd:
  Version:          v1.2.13
  GitCommit:        7ad184331fa3e55e52b890ea95e65ba581ae3429
 runc:
  Version:          1.0.0-rc10
  GitCommit:        dc9208a3303feef5b3839f4323d9beb36df0a9dd
 docker-init:
  Version:          0.18.0
  GitCommit:        fec3683
(base) NDLMAC-49689:~ 49689$ ▮
```

Figure 4-7. *Docker server info*

On the other hand, docker server does all the images, network, volume, and container-related tasks. Basically, it does all the heavy lifting behind the scenes. It looks at the images if present locally; otherwise, it downloads them from Docker Hub. To understand this more, let's look at an example. If we need to run the Nginx server, the command that we would run is docker container run nginx.

A series of steps is followed in order to run the Nginx server using Docker. The moment we issue the command to Docker using the CLI, it gets communicated to the Docker server, which looks for that particular image on the local system. If the image is available locally, it initializes and runs the container with that particular image or else it reaches out to Docker Hub and pulls the particular image (in this case the Nginx image), as shown in Figure 4-8.

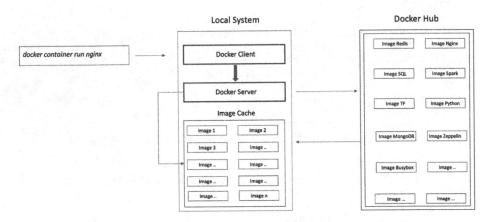

Figure 4-8. *Docker image pull*

It doesn't have to pull the same image twice as once the image gets saved in the local image cache, it gets reused the next time a container is run using that image. In the example to run node.js, we simply have to build the image from Dockerfile, as shown in Figure 4-9.

```
(base) NDLMAC-49689:docker 49689$ docker build -t node .
Sending build context to Docker daemon  2.048kB
Step 1/6 : FROM node:alpine
alpine: Pulling from library/node
cbdbe7a5bc2a: Pull complete
4b0cddaf9d69: Pull complete
63fc8294ef6b: Pull complete
7d5b7d51bbb9: Pull complete
Digest: sha256:7d1366f697f6e906d0cfd9caf1eba6ad3380e20d950b30df9b6e9aca3b141efe
Status: Downloaded newer image for node:alpine
 ---> 72eea7c426fc
Step 2/6 : WORKDIR /app
 ---> Running in c3c4089c0001
Removing intermediate container c3c4089c0001
 ---> a1a967c66fdb
Step 3/6 : Add . /app
 ---> 39a7a9edbca7
Step 4/6 : RUN yarn install
 ---> Running in bc33f6bdec60
yarn install v1.22.4
info No lockfile found.
[1/4] Resolving packages...
[2/4] Fetching packages...
[3/4] Linking dependencies...
[4/4] Building fresh packages...
success Saved lockfile.
Done in 0.09s.
Removing intermediate container bc33f6bdec60
 ---> e596bf09f321
Step 5/6 : EXPOSE 3000
 ---> Running in a0bff6a1308e
Removing intermediate container a0bff6a1308e
 ---> fb3c9e346f8f
Step 6/6 : CMD ["yarn", "start"]
 ---> Running in e3556b39dff2
Removing intermediate container e3556b39dff2
 ---> b099f7feef9a
Successfully built b099f7feef9a
Successfully tagged node:latest
```

Figure 4-9. *Docker image from Dockerfile*

We saw the process to build a Docker image from the Dockerfile. It essentially contains the information in which the container should operate. Docker images are like a snapshot of the file system—a set of directories and files that the application needs to run. They also contain the initialization command to be executed, while containers' instances run using the image. This might be optional as well because we can also override the default startup command by providing a new command during runtime. So, the image is the core file that contains all the key requirements and configuration information pertaining to the application that needs to be run.

Needless to say, it uses the minimum set of configurations required to make the application run, and this makes Docker fast and easy to use, but sometimes the application can be quite complicated in nature, and we end up creating different Docker images for respective components of the application to make them talk to each other for successful execution. As mentioned earlier, the image acts like a source of all the required files and directories to run the specific application. If the image doesn't contain the relevant configuration file or dependencies properly, we cannot run the application using Docker as it depends on these files being executed. For example, say we try to override the default command of the Nginx Docker container and run the ls command instead. As shown in Figure 4-10, we see a bunch of folders already present in this container. These folders are the default configuration files and dependencies required to run the Nginx server. These are part of the Nginx Docker base image to ensure a successful run of an instance of this image anywhere.

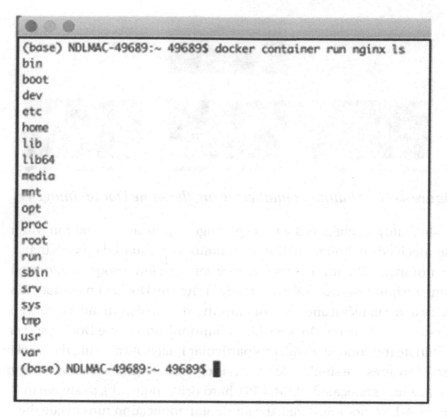

Figure 4-10. *Accessing the running container*

Docker Container

Now that we understand what an image is and what Docker Image all contains, we can move on to containers. The image is used to spin up the containers that are nothing but the running instances of specific images. There can be multiple containers created from the same image to have multiple instances running of the same application, as shown in Figure 4-11.

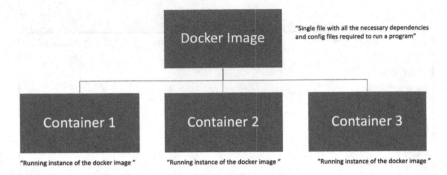

Figure 4-11. *Multiple containers using the same Docker image*

In reality, containers are these packaged applications that run within the specified environment. Docker containers get started up with that environment. The way it does this is by getting a few resources allocated from the host resources. As we already know, the Docker image acts as a file system snapshot and also contains the startup command. Once we pass this image to the Docker CLI, it communicates to the Docker server to initiate the container with this particular image. As a result, the Docker server ensures the similar file system is created inside the container within the resources allocated (RAM, CPU, hard drive, network), as shown in Figure 4-12. Once initialized, the particular application runs inside the container like any other application on the host system.

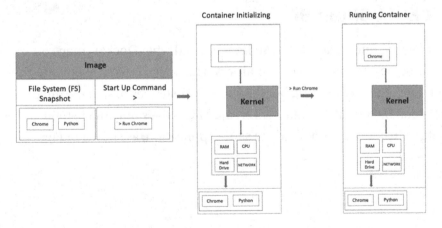

Figure 4-12. *Image to container*

As mentioned, the container is a process or set of processes that makes use of a particular set of resources assigned to it to run the specific application. There are actually two parts to running a container.

- Creating a container

- Running a container

For the first part, we take the base image and get the different resources allocated to the container based upon the default file system and config files of the image. Remember, we don't execute any sort of startup or override command to run the container, as shown in Figure 4-13.

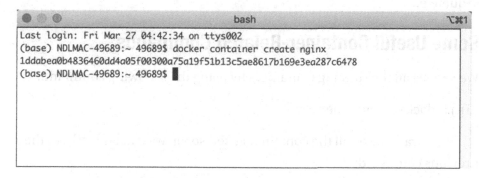

Figure 4-13. *Docker container initialization*

In the second part, once the resources are allocated, running the container includes the execution of the startup command or override command to successfully run the application. The -a indicates the attaching of the container ID to see the output after execution, as shown in Figure 4-14.

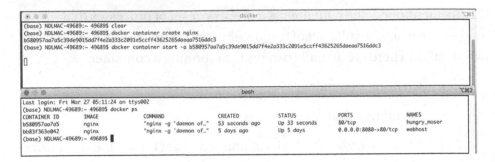

Figure 4-14. *Running a Docker container*

The docker run command executes the previous two stages in sequence.

Some Useful Container-Related Commands

We can list all the running containers by using the following command:

[In]: docker container ps

If we want to see all the containers used so far, we can add –all to the previous command:

[In]: docker container ps –all

If we want to pull a particular image from Docker Hub and use it to customize something based on our specific requirements, we can use the pull command.

[In]: docker pull redis

To run the container using a particular image, we make use of the docker run command.

[In]: docker run

As we have already seen in this example, we can use the build command to create a Docker image from the Dockerfile.

```
[In]: docker build
```

If we want to push the Docker image to the Docker Hub platform, we can use the docker push command with the unique name of the image.

```
[In]: docker push
```

If we want to stop a running container, we can use the docker stop command. If a container doesn't stop within ten seconds, the docker kill command would be activated and the container would be removed instantly.

```
[In]: docker stop
```

If we want to remove a container, we can use the rm command to remove the specific container using a container ID.

```
[In]: docker rm
```

If we want to remove a specific Docker image, we can use the rmi command. Sometimes, we might have to use docker rmi -force to forcefully delete the image.

```
[In]: docker rmi
```

The exec command helps us to get inside a running container.

```
[In]: docker exec
```

The system prune command cleans up your system by removing all the stopped containers and dangling images and frees up a lot of space.

```
[In]: docker system prune
```

Machine Learning Using Docker

In this section, we will build a classification model from scratch and try to deploy it using a Flask app. The only difference from the previous chapter will be to Dockerize the entire application and run it on any platform. We will also make use of the library Flasgger to handle the UI part of the app to make it more intuitive to consume the results. To start with, let's create a new folder in the local system called docker_app.

We will execute this process by following these steps:

1. Train the ML model.

2. Save and export the ML model.

3. Create a Flask app including the UI layer.

4. Build a custom Docker image for the app.

5. Run the app using a Docker container.

6. Stop the container.

Step 1: Training the Machine Learning Model

This is the first stage where we will train the machine learning model on a particular dataset. Since the overall idea is to deploy the ML app, the focus is on the containerizing the app instead of improving the accuracy of the model. Hence, we will not cover too much of the feature engineering or complex ML model. Rather, a simple logistic regression or random forest model will suffice for this exercise. Having said that, we can always replace the given model with a different and better model without impacting the rest of the flow.

The dataset that we have in this case is from an online platform about the historical transactions of customers. It contains the data points such as age, total pages viewed, and whether the customer is a new or repeat customer. The output variable contains whether the customer bought the product online or not. So, we are going to train a simple logistic regression model to make the predictions on the test data and later export it for deployment purposes.

We start by importing the two basic libraries numpy and pandas and later explore the dataset.

```
[In]: import numpy as np
[In]: import pandas as pd

[In]: df = pd.read_csv('online_sales.csv')
[In]: df.shape
[Out]:(316200, 4)
```

We have around 0.3 million records in the dataset with 4 columns in total. Let's see how the data looks and the class balance.

```
[In]:df.head()
[Out]:
```

	age	new_user	total_pages_visited	converted
0	25	1	1	0
1	23	1	5	0
2	28	1	4	0
3	39	1	5	0
4	30	1	6	0

```
[In]: df.converted.value_counts()
[Out]:
0    306000
1     10200
```

We can clearly see there is a skewed target class in this dataset that typically needs to be treated by some undersampling/oversampling technique. However, we are not looking to find the best model accuracy for this exercise, so we will use all the data that we have to build the model. We can also check whether the data contains any missing values to be treated before the model training.

```
[In]: df.info()
[Out]:

        RangeIndex: 316200 entries, 0 to 316199
        Data columns (total 4 columns):
         #   Column                Non-Null Count    Dtype
        ---  ------                --------------    -----
         0   age                   316200 non-null   int64
         1   new_user              316200 non-null   int64
         2   total_pages_visited   316200 non-null   int64
         3   converted             316200 non-null   int64
        dtypes: int64(4)
        memory usage: 9.6 MB
```

As we don't have any missing values in the dataset for any of the columns, we can proceed with splitting the data into training and test sets.

```
[In]: input_columns = [column for column in df.columns if
      column != 'converted']

[In]: print (input_columns)
[Out]: ['age', 'new_user', 'total_pages_visited']

[In]: output_column = 'converted'
[In]:print (output_column)
[Out]: converted

[In]: X = df.loc[:,input_columns].values
[In]: y = df.loc[:,output_column]
```

```
[In]: print (X.shape, y.shape)
[Out]: (316200, 3) (316200,)

[In]: from sklearn.model_selection import train_test_split
[In]: from sklearn.linear_model import LogisticRegression

[In]:X_train,X_test,y_train,y_test=train_test_split(X,y,test_
    size=0.3, random_state=555, stratify=y)

[In]: print(np.sum(y_train))
[In]: 7140

[In]: print(np.sum(y_test))
[Out]: 3060
```

Now that we have our train and test datasets formed, we can move ahead with model training.

Note   In ideal ML scenarios, proper data exploration and feature engineering are advised before model training.

```
[In]:logreg=LogisticRegression(class_weight='balanced').fit(X_
    train,y_train)
[In]: logreg.score(X_test, y_test)
[Out]: 0.93

[In]: predictions=logreg.predict(X_test)

[In]: from sklearn.metrics import confusion_matrix
[In]: from sklearn.metrics import classification_report
```

```
[In]:print(classification_report(y_test,predictions,target_
    names=["Non Converted", "Converted"]))
```

```
[Out]:
                    precision    recall  f1-score   support

    Non Converted        1.00      0.94      0.97     91800
        Converted        0.33      0.92      0.49      3060

         accuracy                            0.94     94860
        macro avg        0.66      0.93      0.73     94860
     weighted avg        0.98      0.94      0.95     94860
```

As we can observe, the overall accuracy of the model is not that bad, and the recall is also good. The precision can be improved further depending on the specific needs. This concludes step 1 for us in the overall process of running ML using Docker.

Step 2: Exporting the Trained Model

Now that we have our model trained, we need to save it using pickle/joblib in order to reuse it later during predictions. In this exercise, we are going to see two kinds of predictions using the ML model.

- Single-customer prediction

- Group prediction (multiple customers)

We have a separate test dataset that can predict for a group of customers instead of only a single customer. Let's see how the results look when we run the app for predictions using Docker.

```
[In]: import pickle
[In]: pickle_out = open("logreg.pkl","wb")
[In]: pickle.dump(logreg, pickle_out)
```

```
[In]: pickle_out.close()
```

Let's load the trained model to see whether it is predicting well on the test data as well as on single-customer values.

```
[In]: pickle_in = open("logreg.pkl","rb")
[In]: model=pickle.load(pickle_in)
```

Here is the prediction for a single customer:

```
[In]: model.predict([[45,0,5]])[0]
[Out]: 0
```

Here is the prediction on a group of customers:

```
[In]: df_test=pd.read_csv('test_data.csv')
[In]: predictions=model.predict(df_test)
[In]: print(list(predictions))
[Out]: [0, 0, 1, 0, 0, 0, 0, 0, 0, 1, 0, 0, 0, 0, 1, 0, 1, 0,
0, 0, 0, 0, 0, 0, 0, 0, 0, 0, 0, 0, 0, 0, 0, 0, 0, 0, 0, 0, 1,
0, 0, 0, 0, 0, 0, 0, 0, 0, 0, 0]
```

As we can observe, the model seems to be making predictions for a single customer as well as a group of customers. Now we can move on to the next step of building a Flask app to run this model.

Step 3: Creating a Flask App Including UI

In the previous chapter, we covered how we can use the Flask framework to deploy an ML model easily. Similarly, in this case, we will build a Flask app along with Flasgger. We start by importing the required libraries to run the Flask app.

```
[In]: from flask import Flask, request
[In]: import numpy as np
```

```
[In]: import pickle
[In]: import pandas as pd
[In]: import flasgger
[In]: from flasgger import Swagger
```

We create the Flask app and wrap it around Swagger. It does all the heavy lifting to represent the result in a well-laid-out manner. This saves a lot of time and effort that might go into writing HTML and CSS.

```
[In]: app=Flask(__name__)
[In]: Swagger(app)
```

Next, we load the trained model for the predictions. Since we are going to make two types of predictions (single customer as well as group of customers), we will need to create two separate functions in our app using a get request as well as a post request. For single-customer predictions, we will use a get request since we will fetch the input values from users, whereas for group prediction, we will use a post request to be able to post a test dataset for predictions. One key thing to remember here is to provide the parameter information to the app regarding the features used in the model to make the predictions properly. The final part is to provide the localhost IP address to run this app.

```
[In]: pickle_in = open("logreg.pkl","rb")
[In]: model=pickle.load(pickle_in)

[In]: @app.route('/predict',methods=["Get"])
[In]: def predict_class():

    """"Predict if Customer would buy the product or not.
    ---
    parameters:
      - name: age
        in: query
```

```
          type: number
          required: true
        - name: new_user
          in: query
          type: number
          required: true
        - name: total_pages_visited
          in: query
          type: number
          required: true

    responses:
        500:
            description: Prediction
    """
    age=int(request.args.get("age"))
new_user=int(request.args.get("new_user"))
total_pages_visited=int(request.args.get("total_pages_
visited"))
    prediction=model.predict([[age,new_user,total_pages_
visited]])
print(prediction[0])
return "Model prediction is"+str(prediction)

[In]: @app.route('/predict_file',methods=["POST"])
[In]: def prediction_test_file():

    """Prediction on multiple input test file.
    ---
    parameters:
      - name: file
        in: formData
```

```
          type: file
          required: true

    responses:
        500:
            description: Test file Prediction
    """
df_test=pd.read_csv(request.files.get("file"))
    prediction=model.predict(df_test)

    return str(list(prediction))

[In]: if __name__=='__main__':
              app.run(debug=True,host='0.0.0.0')
```

This concludes step 3 of creating a Flask app along with the UI layer. We are now ready to get into Docker and build an image to run this app.

Step 4: Building the Docker Image

We are now going to create a Dockerfile and mention all the steps to run this app using Docker. We start with providing the base image first to the Docker server that needs to be pulled from Docker Hub (if not present locally already).

```
FROM continuumio/anaconda3:4.4.0
```

In the next step, we copy all the files and content from the local directory (where we build our ML model and Flask app) to the Docker directory (we can create a new or current Docker directory as well). In this case, we create a new directory called usr/ML/app.

```
COPY . /usr/ML/app
```

The next command is to expose port 5000 of Docker to run this application. Basically, when the request comes from the user, the host system will route this request to port 5000 of Docker where the app will be running. We will also have to do explicit port mapping between the host system and the Docker container while running the container to route the requests properly and access the predictions.

```
EXPOSE 5000
```

The next command is to change Docker's current working directory to the directory where we have copied all the files from the local system (Flask/ML model/data, etc.).

```
WORKDIR /usr/ML/app
```

The next step is to install all the dependencies and required libraries in order to run the app successfully. We have a requirement.txt file that contains the required libraries along with versions.

```
RUN pip install -r requirements.txt
```

The last command in the Dockerfile is always the startup command, and in this case, we want Docker to run the Flask app that we built in the previous step.

```
CMD python flask_api.py
```

This list of commands completes our Dockerfile, and it's now ready to be converted into a Docker image and run as a container, which is the next step.

Step 5: Running the Docker Container

In this step of the process, we build the Docker custom image from the Dockerfile created in the previous step and run the container. We have to go to the terminal in the same directory where all the files are present. In the terminal, we run the docker build command to build the Docker image from Dockerfile.

```
[In]: docker build -t ml_app_docker .
```

The image can be tagged to a specific name to help in running the container. In this case, we tag it to ml_app_docker. Once the Docker image starts to build, there would be a series of steps taking place inside Docker. The first step is that the Docker server would look for the base image. If the base image is not available in the local system, the Docker server would pull it from Docker Hub and might take some time based on the image size. As shown in Figure 4-15, the base image is getting pulled from Docker Hub.

```
Sending build context to Docker daemon  2.955MB
Step 1/6 : FROM continuumio/anaconda3:4.4.0
4.4.0: Pulling from continuumio/anaconda3
8ad8b3f87b37: Downloading [===================================>         ]  35.02MB/51.37MB
fb691515f399: Downloading [===================================>         ]  59.55MB/82.74MB
6c3051db0635: Downloading [==>                                          ]  32.13MB/620.1MB
66faddd8f0d6: Waiting
```

Figure 4-15. *Building the Docker image from the Dockerfile*

Once the base image gets pulled completely, Docker creates a temporary container with that base image. The reason that it's a temporary container is that Docker applies the next step and builds a new container to meet the next requirement. As shown in Figure 4-16, we can observe that at the end of every stage, Docker shows us a new container ID, and after running the last command in the Dockerfile, it provides us with the final container ID.

```
Step 1/6 : FROM continuumio/anaconda3:4.4.0
4.4.0: Pulling from continuumio/anaconda3
8ad8b3f87b37: Pull complete
fb691515f399: Pull complete
6c3051db0635: Pull complete
66faddd8f0d6: Pull complete
Digest: sha256:c6bb52bffe028b4b436b085afa4044db9b3d687a95468c92578467c9c2d4ac31
Status: Downloaded newer image for continuumio/anaconda3:4.4.0
 ---> 795ad88c47ff
Step 2/6 : COPY . /usr/ML/app
 ---> c06ff2ea7ece
Step 3/6 : EXPOSE 5000
 ---> Running in 678c7a20442f
Removing intermediate container 678c7a20442f
 ---> f7ee7a1be45e
Step 4/6 : WORKDIR /usr/ML/app
 ---> Running in 91ec7d7aeb98
Removing intermediate container 91ec7d7aeb98
 ---> 231143327a32
Step 5/6 : RUN pip install -r requirements.txt
```

Figure 4-16. *Docker server info*

The `requirements.txt` file contains all the dependencies and libraries, and hence Docker installs all those as well, as shown in Figure 4-17.

```
Collecting Flask==1.1.1 (from -r requirements.txt (line 1))
  Downloading https://files.pythonhosted.org/packages/9b/93/628509b8d5dc749656a9641f4caf13540e2cdec85276964ff8f43bbb1d3b/Flask-1.1.1-py2.py3-none-any.whl (94kB)
Collecting gunicorn==19.9.0 (from -r requirements.txt (line 2))
  Downloading https://files.pythonhosted.org/packages/8c/da/b8dd8deb741bff556db53902d4706774c8e1e67265f69528c14c003644e6/gunicorn-19.9.0-py2.py3-none-any.whl (112kB)
Collecting itsdangerous==1.1.0 (from -r requirements.txt (line 3))
  Downloading https://files.pythonhosted.org/packages/76/ae/44b03b253d6fade317f32c24d100b3b35c2239807046a4c953c7b89fa49e/itsdangerous-1.1.0-py2.py3-none-any.whl
Collecting Jinja2==2.10.1 (from -r requirements.txt (line 4))
  Downloading https://files.pythonhosted.org/packages/1d/e7/fd8b501e7a6dfe492a433deb7b9d833d39ca74916fa8bc63dd1a4947a671/Jinja2-2.10.1-py2.py3-none-any.whl (124kB)
Collecting MarkupSafe==1.1.1 (from -r requirements.txt (line 5))
  Downloading https://files.pythonhosted.org/packages/b2/5f/23e0023be6bb885d00ffbefad2942bc51a620328ee910f64abe5a8d18dd1/MarkupSafe-1.1.1-cp36-cp36m-manylinux1_x86_64.whl
Collecting Werkzeug==0.15.5 (from -r requirements.txt (line 6))
  Downloading https://files.pythonhosted.org/packages/d1/ab/d3bed6b92042622d24decc7aadc8877badf18aeca1571045840ad4956d3f/Werkzeug-0.15.5-py2.py3-none-any.whl (328kB)
Requirement already satisfied: numpy>=1.9.2 in /opt/conda/lib/python3.6/site-packages (from -r requirements.txt (line 7))
Requirement already satisfied: scipy>=0.15.1 in /opt/conda/lib/python3.6/site-packages (from -r requirements.txt (line 8))
Collecting scikit-learn==0.22.1 (from -r requirements.txt (line 9))
  Downloading https://files.pythonhosted.org/packages/d1/48/e9fa9e252abcd1447eff6f9257636af31758a6e46fd5ce5d3c879f6907cb/scikit_learn-0.22.1-cp36-cp36m-manylinux1_x86_64.whl (7.0MB)
Requirement already satisfied: matplotlib>=1.4.3 in /opt/conda/lib/python3.6/site-packages (from -r requirements.txt (line 10))
Requirement already satisfied: pandas>=0.19 in /opt/conda/lib/python3.6/site-packages (from -r requirements.txt (line 11))
Collecting flasgger==0.9.4 (from -r requirements.txt (line 12))
  Downloading https://files.pythonhosted.org/packages/41/d6/c06817828fc6ef724d6ac60f3f558e048300c5ffd038b60d289e6599c117/flasgger-0.9.4-py2.py3-none-any.whl (3.8MB)
```

Figure 4-17. *Installing dependencies*

Once all the commands in the Dockerfile get executed and we have the final image built from the Dockerfile, as shown in Figure 4-18, we can initiate the container to run our ML app.

```
Removing intermediate container 6601f5ed213c
 ---> a2a2facb1459
Step 6/6 : CMD python flask_api.py
 ---> Running in 9b88041066f3
Removing intermediate container 9b88041066f3
 ---> 7b3b71263bd2
Successfully built 7b3b71263bd2
Successfully tagged ml_app_docker:latest
```

Figure 4-18. *Final Docker image*

The key thing to remember here is to do the explicit port mapping in order to route the requests from the host to the Docker port. We make use of `-p` and map host port 5000 to the Docker container port 5000.

[In]: docker container run -p 5000:5000 ml_app_docker

[Out]:

As shown in Figure 4-19, the app starts up successfully, and everything is running inside a Docker container. To access the app, we simply have to go to `http://localhost://5000/apidocs` to load the Swagger UI page, as shown in Figure 4-20.

```
* Serving Flask app "flask_api" (lazy loading)
* Environment: production
  WARNING: This is a development server. Do not use it in a production deployment.
  Use a production WSGI server instead.
* Debug mode: on
* Running on http://0.0.0.0:5000/ (Press CTRL+C to quit)
* Restarting with stat
* Debugger is active!
* Debugger PIN: 290-890-899
```

Figure 4-19. *Access the running ml-app*

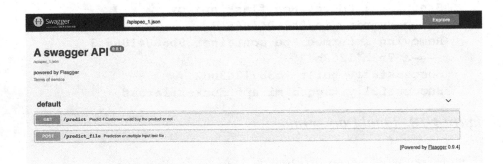

Figure 4-20. *Swagger API to access ml-app*

Two tabs are present in the app.

- Get

- Post

The prediction based on get requests is applicable for single-customer predictions, whereas the Post tab is for the test data prediction (customer group). Once we click the Get tab, we can see the options to provide input parameters on which the prediction needs to be made. These are the same parameters that we used in the model training as well as called in the Flask app code. The top-right corner contains a "Try it out" tab that allows us to fill in the values for the input parameters, as shown in Figure 4-21.

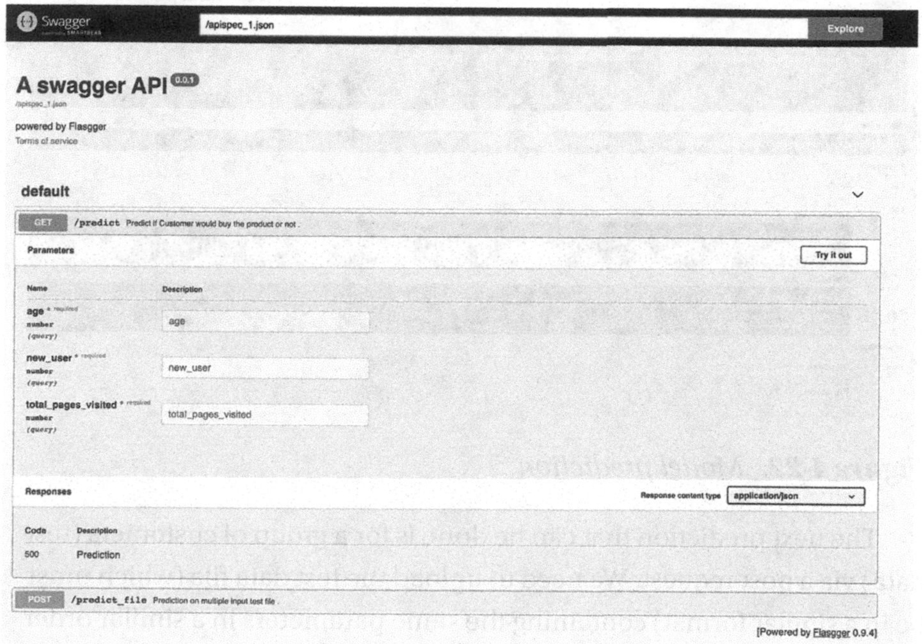

Figure 4-21. *Get request*

We can fill in the values for all three parameters for a test customer, as shown in Figure 4-21, and click the Execute tab. Upon the execution call, the request goes to the app, and predictions are made by the model. The result of the model prediction is displayed in the Prediction section of the page. It also provides alternative ways to access the result such as using a curl command or URL, as shown in Figure 4-22.

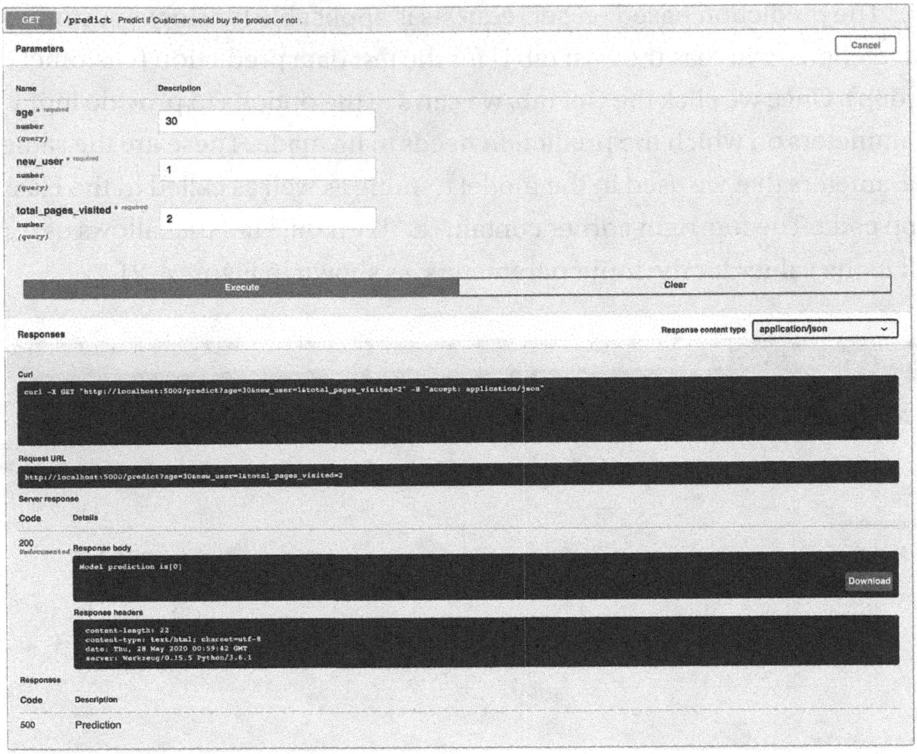

Figure 4-22. *Model prediction*

The next prediction that can be done is for a group of customers (test data) via a post request. We need to upload the test data file (which must be in a similar format) containing the same parameters in a similar order as shown in Figure 4-23. The model would make the prediction, and the results would be displayed upon execute, as shown in Figure 4-24.

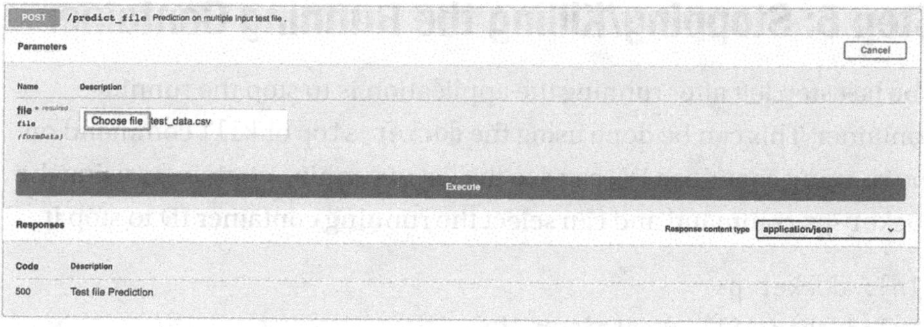

Figure 4-23. *Uploading test data*

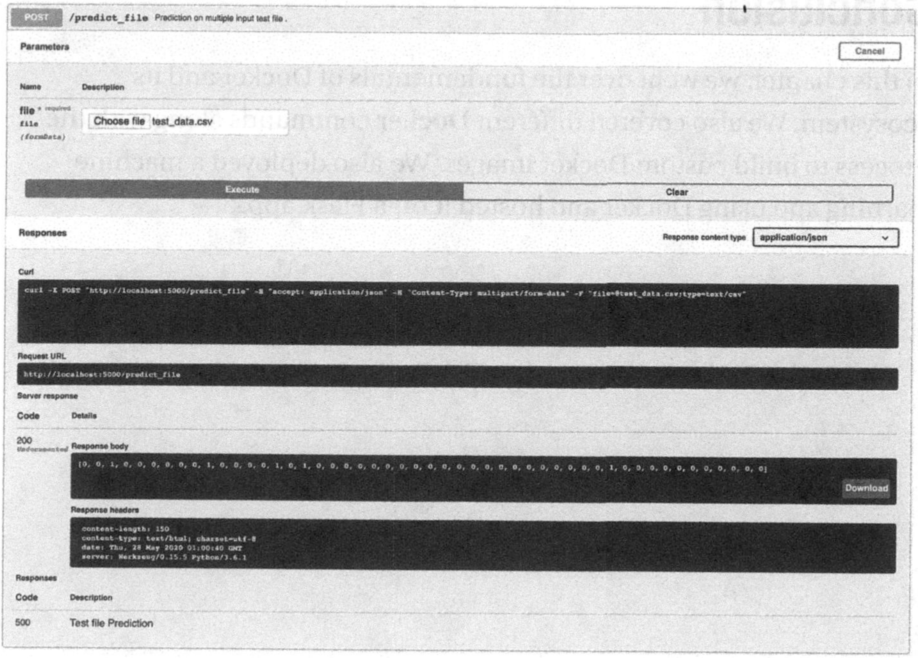

Figure 4-24. *Multiple predictions*

Step 6: Stopping/Killing the Running Container

The last step left after running the application is to stop the running
container. This can be done using the `docker stop` or `kill` command on
the running container. We can see the list of running containers using the
`docker ps` command and can select the running container ID to stop it.

```
[In]: docker ps
[In]:docker kill <Container_ID>
```

Conclusion

In this chapter, we went over the fundamentals of Docker and its
ecosystem. We also covered different Docker commands along with the
process to build custom Docker images. We also deployed a machine
learning app using Docker and hosted it on a Flask app.

CHAPTER 5

Machine Learning Deployment Using Kubernetes

In the previous chapter, we saw how we can containerize an app and deploy it using Flask. In this chapter, we will deploy the same ML app using an orchestration platform (Kubernetes). This chapter covers two main topics. It first covers the basics of the Kubernetes platform and how it handles deployments. Then the chapter covers how to deploy the ML app using Kubernetes. More specifically, the chapter covers the following:

- What is Kubernetes
- Google Cloud Platform
- ML model deployment using Kubernetes

Kubernetes is an open source container orchestration engine to deploy and manage large containers. It comes with multiple levers to manage scheduling, clusters, deployments, and load balancing, and it has many more capabilities in terms of running microservices using containers. Kubernetes was created at Google originally and later donated to the Cloud Native Computing Foundation (CNCF). It is now managed and maintained by CNCF and has strong community support and users around the globe.

© Pramod Singh 2021
P. Singh, *Deploy Machine Learning Models to Production*,
https://doi.org/10.1007/978-1-4842-6546-8_5

The prime advantage of using Kubernetes is to be able run a huge number of containers without worrying about the deployments and cluster management details. To give you an example, Google runs roughly 2.5 billion containers using Kubernetes to run its services for users. Although there are other container orchestration platforms such as Docker Swarm and Marathon from Apache Mesos, Kubernetes has quickly become the default COE for users because of a couple of reasons. First, it has been tried and tested for extreme cases (since it was developed at Google). Second, it has a rich set of features and an underlying sophisticated architecture that allows it to run highly scalable services.

Kubernetes is available on different cloud platforms such as Google Cloud Platform's Google Kubernetes Engine (GKE), AWS EC2 Container Service, and Microsoft Azure Container. In this chapter, we are going to make use of GKE to deploy the ML app. A similar approach can be used to deploy it on AWS and Azure, but Google provides its users with some free credits that can be easily utilized to complete this deployment.

Kubernetes Architecture

Kubernetes under the hood works like any other distributed applications. It works on the master-worker node principle. The master node issues the commands/instructions to the worker nodes that are responsible for computing and executing the tasks. In the case of Kubernetes, there can be more than a single master as well. The worker nodes are typically the virtual machines that are used to compute and store the data. If we look at the architecture of Kubernetes, as shown in Figure 5-1, we can observe that there are two main parts to it.

- Master
- Worker

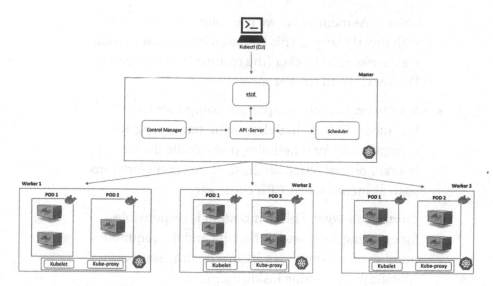

Figure 5-1. *Kubernetes architecture*

The CLI tool that is used to interact with the Kubernetes object is known as kubectl.

Kubernetes Master

A Kubernetes *master* is responsible for running the entire cluster and ensuring the scheduling and provisioning of the pods (see the "Worker Nodes" section). Like any other master of a distributed application, it's responsible for communicating and tracking the status of worker nodes and provisioning a healthy node in the case of worker node failure. Inside the Kubernetes master, there are four main components.

- *API server*: This component allows us to interact with different Kubernetes objects. It validates and configures the API for different objects such as pods, deployments, load balancers, etc. It allows us to add, delete, update, and display the information regarding the Kubernetes objects. As mentioned, we use kubectl to interact with the API server. This is somewhat similar to what we have seen in Docker (the relationship between the Docker CLI and Docker server).

- *Scheduler*: The other important component inside the master is the Scheduler. As the name suggests, it is responsible for scheduling pods inside the cluster (worker nodes) based on the declared configurations such as memory size, CPU cores, etc.

- *Control Manager*: This component is responsible for ensuring the cluster is healthy and the required number of worker nodes are in a healthy state with specific pods running inside them.

- *etcd*: This component captures the current state of the cluster in the form of a key value. It is a distributed lightweight key-value database.

Worker Nodes

The worker nodes are the typical virtual machines in the cloud or the physical servers in a data center. They are responsible for computing and for storing the data of the running application. Each worker node must have a container runtime (Docker/rkt) to be able to run containers inside them. Each worker node is capable of running one or more pods inside them. A pod is the most basic unit in Kubernetes. Pods are essentially the scheduling units in Kubernetes and contain one or more containers. Ideally, each pod should contain one container, but in the case of dependent containers, they can be run inside the same pod. The pod acts as a wrapper around containers that allows us to interact with the containers inside them. There are two additional components inside a worker node that help to communicate with the Kubernetes master.

- kubelet: This is the primary node agent that runs on each worker node. It ensures that the containers running inside the pods are according to the specs submitted to the API server. If any changes are observed or any of the pods are down, it starts a new pod on the same node with new containers as per the configuration information.

- kube-proxy: This component is responsible for maintaining the distributed networking configuration of the cluster. It manages the network configuration for the nodes, pods, and containers running inside the pods and ensures the running services are accessible to the outside world.

ML App Using Kubernetes

Now that we have a basic understanding of the Kubernetes platform, we can move into the second part of this chapter and deploy an ML app using Kubernetes. To deploy our app using Kubernetes, the first step is to push the local code into a Git repo so that we can clone it later. In an ideal scenario, the data is usually stored in a Google storage bucket, but for simplicity purposes we have packaged everything inside a Docker image because we don't have a large dataset. As mentioned, we are going to use GCP to avail of the free credits provided by Google to use cloud resources such as Google Kubernetes Engine.

Google Cloud Platform

Google Cloud Platform is a huge platform containing lots of tools and services for a variety of requirements. It is impossible to cover all the aspects of GCP, and hence we are going to focus on certain services to deploy the ML app. We are going to use a few components such as GKE and Google Container Registry (GCR) for our deployment. The prerequisite to using GCP is to create a Google account and log in to the Google Cloud Platform by visiting `https://console.cloud.google.com/`.

Note   There is a cost associated with every tool and service used in GCP, and hence readers are encouraged to learn more about the billing rates on the Google Console details page. However, for this deployment, the free credits provided by Google will be enough.

Once we log in, the first step is to create a new project in the Google Console. This is how we allocate separate resources for this particular project. We can name this project as per our preference as shown in Figure 5-2. For example, in this case, I have named this new project `ml-model`, as shown in Figure 5-3.

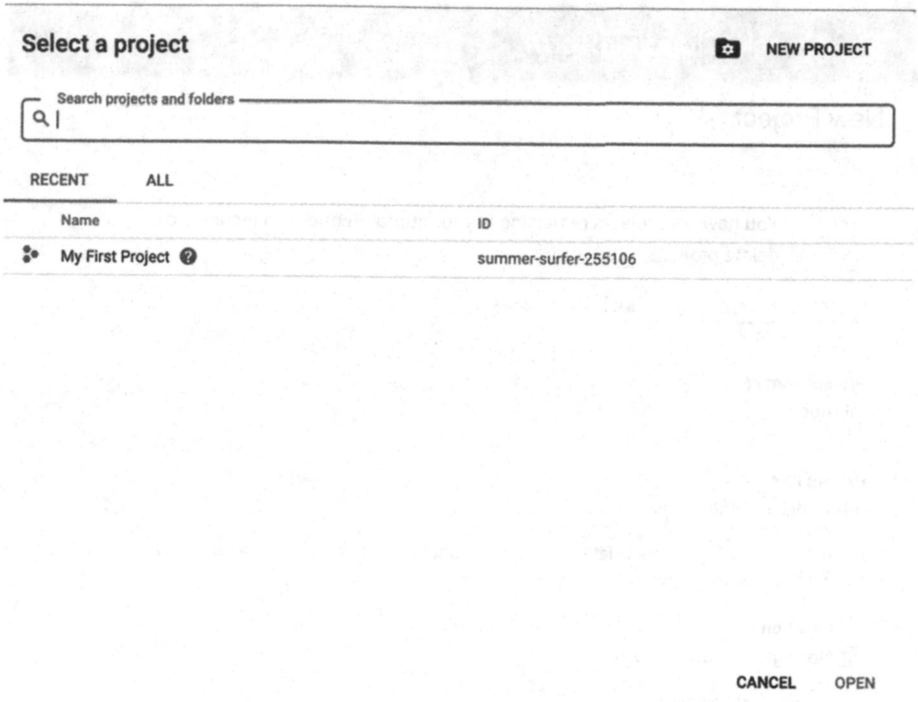

Select a project   ⊡   NEW PROJECT

Search projects and folders
Q |

RECENT   ALL

Name	ID
My First Project ❓	summer-surfer-255106

CANCEL   OPEN

Figure 5-2. *GCP project*

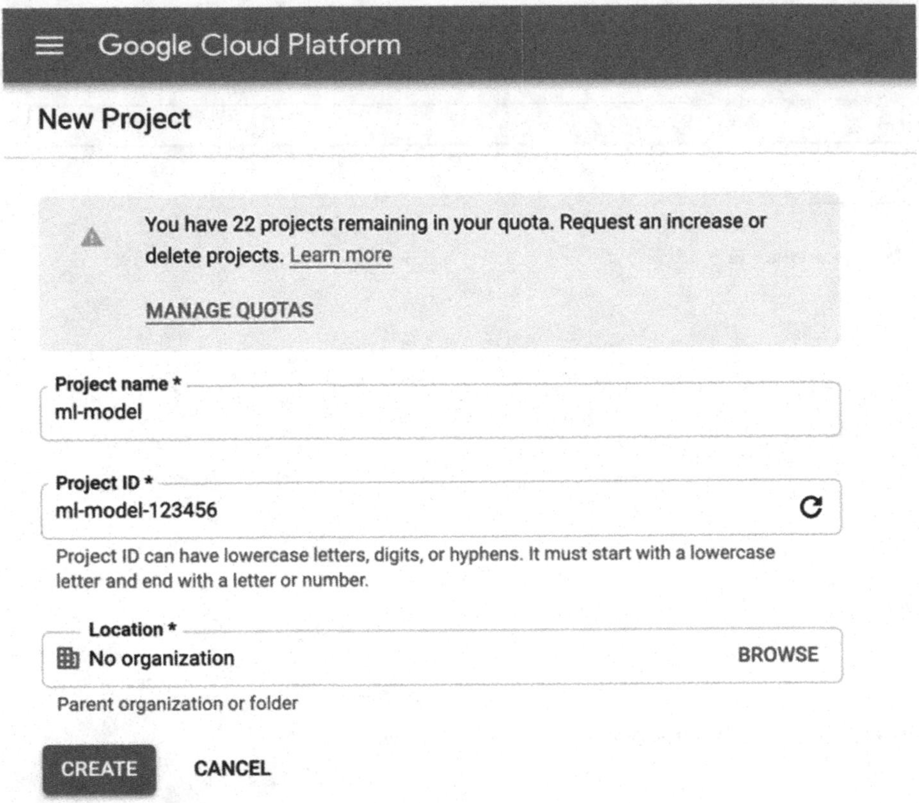

Figure 5-3. *New project in GCP*

Once the new project is created, it will be reflected on the home page on the Project Info tab, as shown in Figure 5-4. It will contain a project name and a project ID. There is also an option to add more people to the project if there are more people working on it.

Figure 5-4. *Project info in GCP*

Once the project is created successfully, we need to enable other services that are required to deploy the app. The first service that we need to select is the Container Registry API. GCP's Container Registry allows us to access Docker images inside the Google project to deploy them on Kubernetes. To do so, we go to the GCP menu tab and select Container Registry, as shown in Figure 5-5. There are two options under Container Registry.

- Images

- Settings

We need to select Images, and by default, if we look inside Images, it should show an empty pane, as shown in Figure 5-6. This is because we have yet to push any image in the Container Registry. You might have previous images present if you have used Container Registry before for some other application, but for new users it should not contain any prior images.

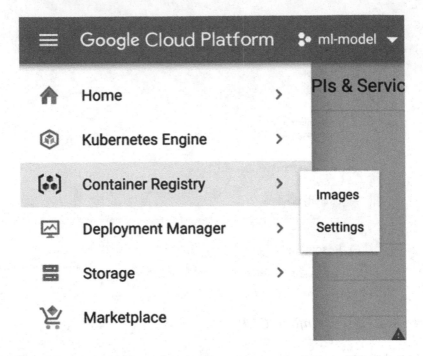

Figure 5-5. *Container Registry menu item in GCP*

To use and push Docker images to the Container Registry, we should enable the Container Registry API by selecting the enable option, as shown in Figure 5-7.

Figure 5-6. *Images in Container Registry*

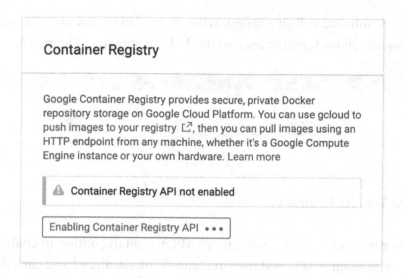

Figure 5-7. *Enabling the Container Registry API*

The next API that we have to enable is for the Kubernetes Engine itself. Select the APIs & Services option on the main menu tab and select Dashboard, as shown in Figure 5-8.

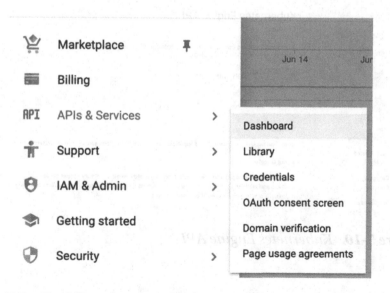

Figure 5-8. *Enabling the APIs & Services item*

137

The dashboard will give access to the entire API library on GCP, and we can now search for Kubernetes-specific APIs, as shown in Figure 5-9.

Figure 5-9. *API Library*

Once we search for the Kubernetes API, we get the option to enable the Kubernetes Engine API, as shown in Figure 5-10. Go ahead and click the enable option; it might take a couple of minutes to completely enable it.

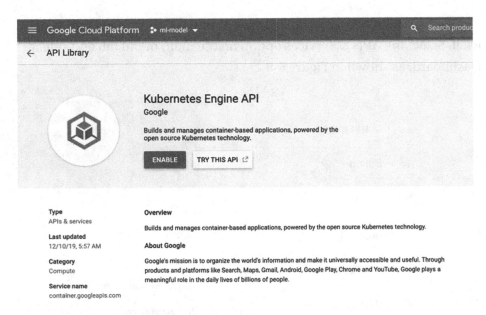

Figure 5-10. *Kubernetes Engine API*

Now that we have created a new project and enabled the required services, we can start the configuration steps inside Google Cloudshell. We can open Google Cloudshell by clicking the terminal icon at the top-right corner of the window, as shown in Figure 5-11.

Figure 5-11. *Enabling Google Cloudshell*

Once the terminal opens, the first step is to clone the Git repo where the source code/data is available. In this case, we clone the docker_flask_ gke.git repo, as shown in Figure 5-12.

```
CLOUD SHELL
Terminal     (ml-model-123456) ×  + ▾

pramodsinghwalmart@cloudshell:~ (ml-model-123456)$ git clone https://github.com/pramodchahar/docker_flask_gke.git
Cloning into 'docker_flask_gke'...
remote: Enumerating objects: 9, done.
remote: Counting objects: 100% (9/9), done.
remote: Compressing objects: 100% (9/9), done.
remote: Total 9 (delta 0), reused 0 (delta 0), pack-reused 0
Unpacking objects: 100% (9/9), done.
pramodsinghwalmart@cloudshell:~ (ml-model-123456)$ ls
docker_flask_gke   README-cloudshell.txt
pramodsinghwalmart@cloudshell:~ (ml-model-123456)$ cd docker_flask_gke/
pramodsinghwalmart@cloudshell:~/docker_flask_gke (ml-model-123456)$ ls
Dockerfile  flask_api.py  logreg.pkl  ML.ipynb  online_sales.csv  requirements.txt  test_data.csv
pramodsinghwalmart@cloudshell:~/docker_flask_gke (ml-model-123456)$ ▮
```

Figure 5-12. *Cloning the app files*

To confirm everything has been set up properly, we go inside the new cloned folder and run a quick ls command and see whether the files and code are available in the Google Cloudshell folder. The next step is to create some of the environment variables to keep the deployment consistent. We declare the project ID variable, which is the project ID that we created at the start. The next step is to build a Docker image using the Dockerfile inside our directory. Since the format is the same as what we used on our local system, we can make use of the same Dockerfile to build a new Docker image. We use the docker build command and tag it with the name that includes the project ID. The build process will take some amount of time (depending upon the Internet bandwidth). It will run the same steps to build the Docker image that we saw in the previous chapter, as shown in Figure 5-13.

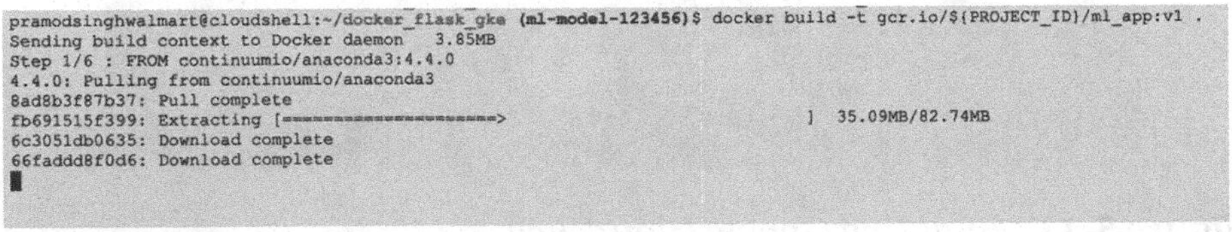

```
pramodsinghwalmart@cloudshell:~/docker_flask_gke (ml-model-123456)$ docker build -t gcr.io/${PROJECT_ID}/ml_app:v1 .
Sending build context to Docker daemon   3.85MB
Step 1/6 : FROM continuumio/anaconda3:4.4.0
4.4.0: Pulling from continuumio/anaconda3
8ad8b3f87b37: Pull complete
fb691515f399: Extracting [=======================>                     ]   35.09MB/82.74MB
6c3051db0635: Download complete
66faddd8f0d6: Download complete
```

Figure 5-13. *Building a Docker image*

[In]: export PROJECT_ID=ml-model-123456
[In]: docker build -t gcr.io/${PROJECT_ID}/ml_app:v1 .

During the build process, it will go over all the necessary steps to install the dependencies that are mentioned in the requirement.txt file, such as setting the working directory, as shown in Figure 5-14.

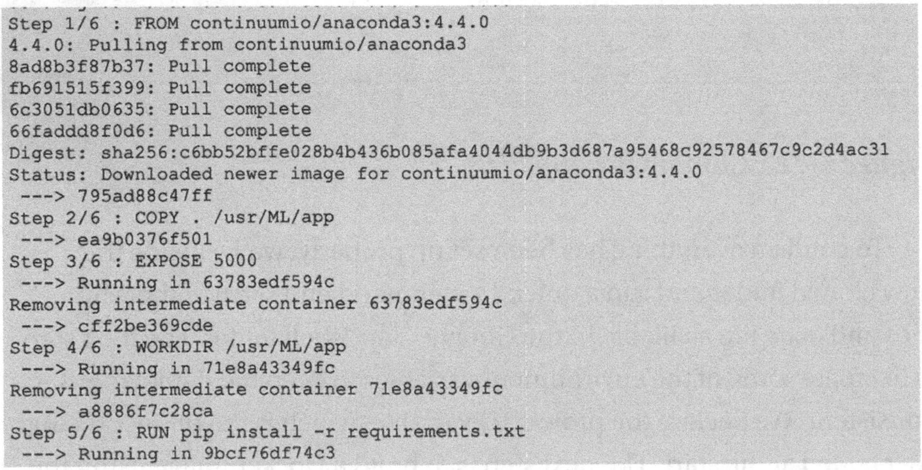

```
Step 1/6 : FROM continuumio/anaconda3:4.4.0
4.4.0: Pulling from continuumio/anaconda3
8ad8b3f87b37: Pull complete
fb691515f399: Pull complete
6c3051db0635: Pull complete
66faddd8f0d6: Pull complete
Digest: sha256:c6bb52bffe028b4b436b085afa4044db9b3d687a95468c92578467c9c2d4ac31
Status: Downloaded newer image for continuumio/anaconda3:4.4.0
 ---> 795ad88c47ff
Step 2/6 : COPY . /usr/ML/app
 ---> ea9b0376f501
Step 3/6 : EXPOSE 5000
 ---> Running in 63783edf594c
Removing intermediate container 63783edf594c
 ---> cff2be369cde
Step 4/6 : WORKDIR /usr/ML/app
 ---> Running in 71e8a43349fc
Removing intermediate container 71e8a43349fc
 ---> a8886f7c28ca
Step 5/6 : RUN pip install -r requirements.txt
 ---> Running in 9bcf76df74c3
```

Figure 5-14. *Executing Dockerfile commands*

We can confirm that the image was successfully built by using the docker images command to list the Docker images; the newly built image should show up, as shown in Figure 5-15 (as there are no previous Docker images for this project).

```
[In]: docker images
[Out]:
```

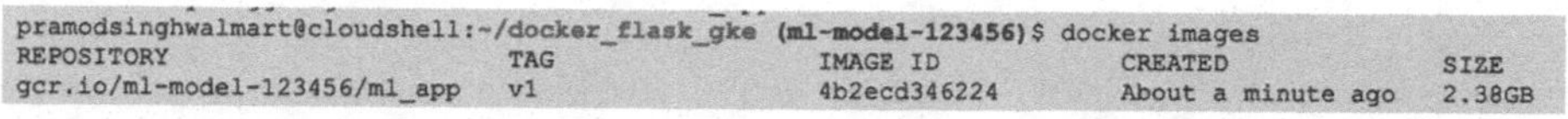

```
pramodsinghwalmart@cloudshell:~/docker_flask_gke (ml-model-123456)$ docker images
REPOSITORY                         TAG          IMAGE ID         CREATED            SIZE
gcr.io/ml-model-123456/ml_app      v1           4b2ecd346224     About a minute ago  2.38GB
```

Figure 5-15. *Docker images*

The next step is to provide Google authentication by using the gcloudauth command and pushing the Docker image created earlier to the Google Container registry. We make use of the docker push command and provide the location of it in the Google Container registry to upload the Docker image, as shown in Figure 5-16.

```
[In]: gcloudauth configure-docker
[In]: docker push gcr.io/${PROJECT_ID}/ml_app:v1
```

```
pramodsinghwalmart@cloudshell:~/docker_flask_gke (ml-model-123456)$ docker push gcr.io/${PROJECT_ID}/ml_app:v1
The push refers to repository [gcr.io/ml-model-123456/ml_app]
ea1c855ba6ed: Pushing [==========================================>]  61.49MB
148cd6f8a127: Pushed
ca173dc10e31: Pushed
54e10c08a841: Pushing [=>                                         ]  66.21MB/1.993GB
1f09b1beaa90: Pushing [==========================>                ]  97.47MB/195.2MB
9e63c5bce458: Pushing [=====>                                     ]  15.68MB/125.1MB
```

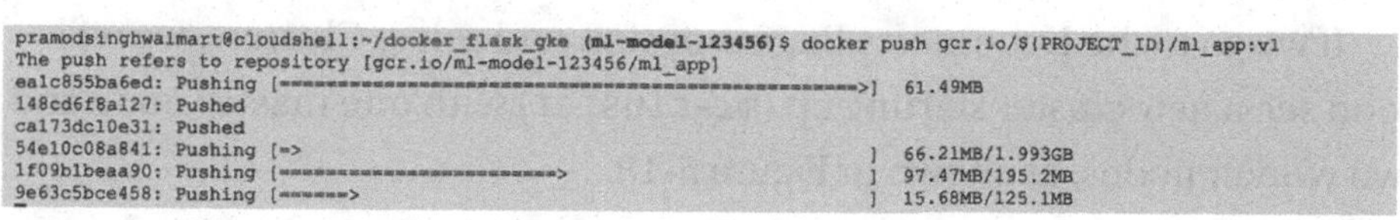

Figure 5-16. *Pushing the Docker image to GCR*

It might take some time based on the image size to push to the Google Container Registry. Once that's completed, we can open the images under the Container Registry, and the Docker image that we uploaded should be present under the ml_app folder, as shown in Figure 5-17.

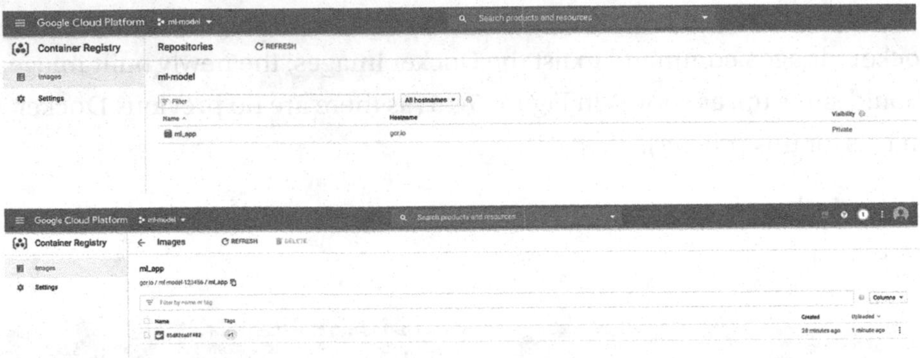

Figure 5-17. *Docker image in GCR*

Now that we have pushed the Docker image to the Container Registry, we can set up other configurations. We set the project to the project ID, we set the compute zone to us-central1, and we create a small two-node cluster called ml-cluster (going for a higher configuration cluster might cost you more).

```
[In]: gcloud config set project $PROJECT_ID
[In]: gcloud config set compute/zone us-central1
[In]: gcloud container clusters create ml-cluster --num-nodes=2
```

If we go to the Kubernetes Engine option and select Clusters, we will soon see a new cluster starting up (ml-cluster) with one master node and two worker nodes, as shown in Figure 5-18.

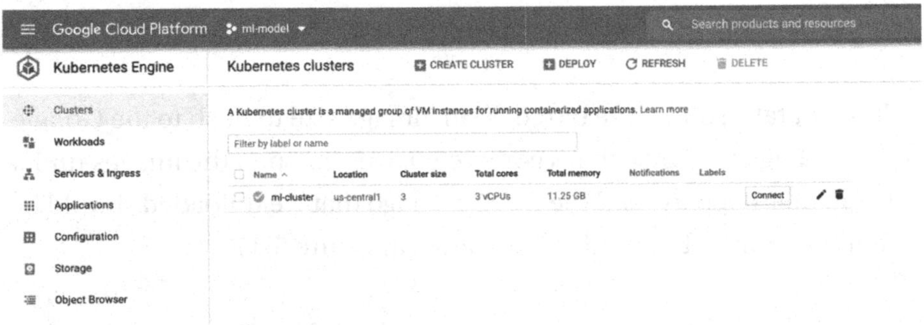

Figure 5-18. *Kubernetes cluster*

Now that the cluster is up and running, we can deploy Docker containers to run the ML app using the base Docker image that we built earlier. We make use of the `create deployment` command and pass the specific image location as input. The other thing that we need to do is to expose the deployed app on port 5000. Once the deployments are complete, we can access the running services by using the `get service` command, as shown in Figure 5-19.

```
[In]: kubectl create deployment ml-app  --image=gcr.
      io/${PROJECT_ID}/ml_app:v1
```

```
[In]: kubectl expose deployment ml-app --type=LoadBalancer
      --port 80 --target-port 5000
```

```
pramodsinghwalmart@cloudshell:~/docker_flask_gke (ml-model-123456)$ kubectl get service
NAME          TYPE            CLUSTER-IP        EXTERNAL-IP       PORT(S)         AGE
kubernetes    ClusterIP       10.51.240.1       <none>           443/TCP          8m6s
ml-app        LoadBalancer    10.51.255.205     34.71.218.107     80:32304/TCP     2m15s
```

Figure 5-19. *Docker image in GCR*

Kubernetes shows the external IP address through which the running app can be accessed. We can simply go to that external IP address and add `apidocs` to it (to access the Swagger API), as shown in Figure 5-20.

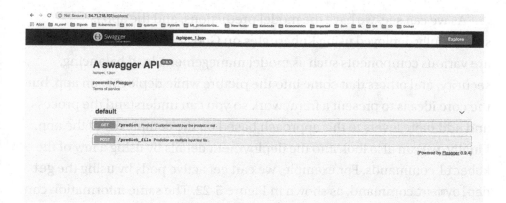

Figure 5-20. *Accessing the app from outside*

We can test whether the app is functioning properly by passing some dummy values as input to the model and clicking the Execute button, as shown in Figure 5-21.

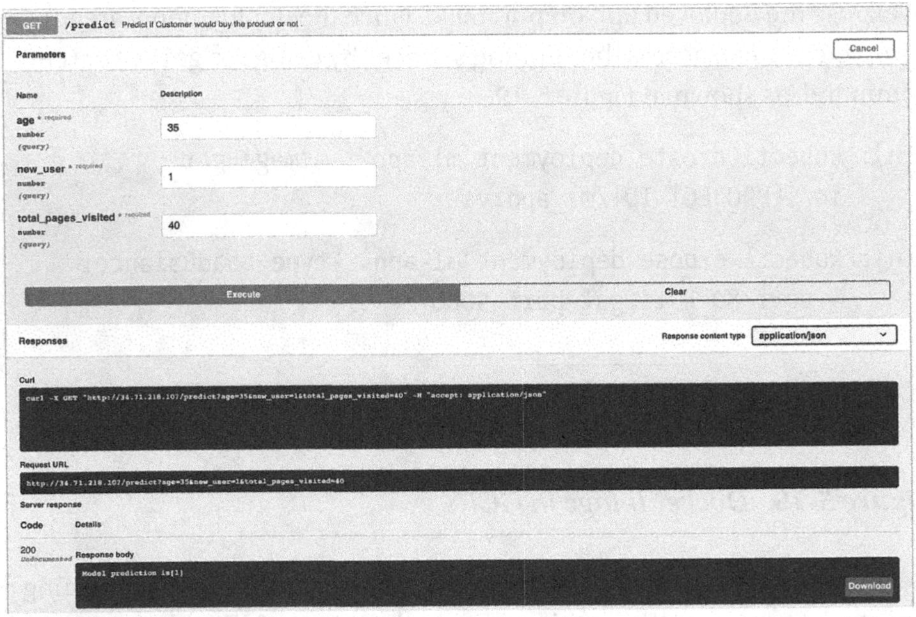

Figure 5-21. *ML app prediction*

As we can see, we have the model predictions, and the app is successfully deployed using Kubernetes on Google Cloud Platform. There are various components such as model management, load balancing, security, and others that come into the picture while deploying an app, but the core idea is to present a framework so you can understand the process and add more levers to this approach based on the complexity of the app. Finally, we can also look into the deployment details by using a few of the kubectl commands. For example, we can get active pods by using the get deployment command, as shown in Figure 5-22. The same information can be viewed on the Workloads tab inside the Kubernetes Engine option, as shown in Figure 5-23.

```
pramodsinghwalmart@cloudshell:~/docker_flask_gke (ml-model-123456)$ kubectl get deployment
NAME      READY   UP-TO-DATE   AVAILABLE   AGE
ml-app    2/2     2            2           7m2s
```

Figure 5-22. *Active pods in Kubernetes*

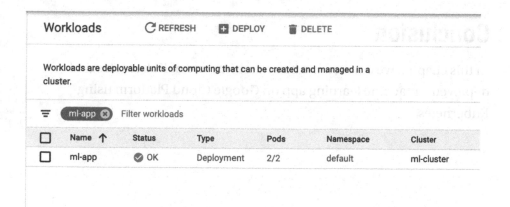

Figure 5-23. *Status of deployed app*

If required, the number of replicas of the running application can
be scaled up or scaled down using the scale deployment command in
Kubernetes; however, you also need to increase the number of nodes in
the cluster to meet the minimum requirements to run that many replicas
of the application. In this case, since we are using small clusters with just
two nodes, we can go up to three to four replicas.

[In]: kubectl scale deployment ml-app --replicas=3

As mentioned previously, all these resources have an associated cost,
and hence it's important for us to delete these used resources to avoid any
costs. To delete the active cluster and remove all the resources, we need to
use the following commands:

[In]: gcloud container clusters delete ml-cluster

145

It is also advisable to delete the project and related files (image, data, etc.) as well once the process is complete to avoid any charges for continued usage of GCP resources.

Conclusion

In this chapter, we covered the basic architecture of Kubernetes and deployed a machine learning app on Google Cloud Platform using Kubernetes.

Index

S, T

U, V, W, X, Y, Z

Printed in the United States
By Bookmasters